SpringerBriefs in Education

For further volumes:
http://www.springer.com/series/8914

Karen Englander

Writing and Publishing Science Research Papers in English

A Global Perspective

Springer

Karen Englander
Department of Languages
Literatures and Linguistics
York University
Toronto, ON
Canada

ISSN 2211-1921 ISSN 2211-193X (electronic)
ISBN 978-94-007-7713-2 ISBN 978-94-007-7714-9 (eBook)
DOI 10.1007/978-94-007-7714-9
Springer Dordrecht Heidelberg NewYork London

Library of Congress Control Number: 2013948858

Printed on acid-free paper

Springer is part of Springer Science+Business Media (www.springer.com)

Preface

This book is intended for scientists and graduate students who need to write and publish scientific research. The contents of this book are based on the most current research published by experts who analyze the papers that scientists write—applied linguists, sociolinguists, geopolitical policy analysts, and bibliometricians, among others. In other words, the research presented here is not based on intuition; it is based on empirical investigations of why and how scientific research is written and published on a global scale. Every effort was made to eliminate technical jargon in order to produce a book that constitutes a practical and sound overview of scientific writing in the world today.

My personal motive for creating this book emerges from the dozen years I have spent working with scientists who need to publish their research in English. Their struggles, victories, and furtherance of their disciplines always inspire me. I feel a great gratitude to those men and women who have entrusted me with their manuscripts and their experiences. It is my great hope that, in some small way, this book can make writing and publishing more accessible and less frustrating to other novices, non-native-speakers of English and established scientists, from this time onward.

This book would not have been created without the initial encouragement of my colleague David Ian Hanauer. Our previous book "Scientific Writing in a Second Language" provided a basis for undertaking this project. I am invariably grateful for his scholarship and his humanity.

Finally, I wish to thank Javad Bourbour Shirazi, graduate student at York University, who has gone well beyond the bounds of all reasonable expectations in proofreading, formatting, and reference checking this entire manuscript. His dedication and scrupulous attention has made this a better book. Any errors, of course, remain mine.

Toronto, July 2013 Karen Englander

Contents

Introduction

This book is designed for scientists and graduate students who wish to understand in a succinct format the current research on the research article. This book provides a comprehensive, state-of-the-art review of the current knowledge concerning writing and publishing scientific research papers and its social contexts. A special aspect of the book is that it deals with English and with non-Anglophone science writers. Research from many diverse disciplines is collected and synthesized to produce a unique contribution to understanding contemporary writing and publishing in science.

The book is divided into three parts. Part I considers "The Context". It discusses the production of scientific articles, and it documents the rise and implications of English as the language of science. The impact factor as a dominating element in determining the status of journals, scientists, and their work is explored. The context for creating science and publishing papers is also examined through the effect of research money and English competence as predictors of publishing success. The role of collaborations, teams, and networks in supporting publication activity is examined.

Part II discusses "The Text" itself. This part reports on how science articles are constructed and how English articles compare with those written in other languages. This largest part of the book brings together empirical studies of the linguistic and rhetorical structure of research papers and presents these findings in reader-friendly terms. Readers of this part can gain insight as to how to better write their own science manuscripts.

Part III reports on "The People" involved in writing and publishing science articles. This part begins with the trajectory of learning to write science in university settings. It discusses the strengths and challenges faced by people with different degrees of science writing expertise, and documents the ways that people with lesser expertise can learn from more established scientists. In turn, techniques for established scientists to support newcomers are discussed. This part ends with a discussion of the role of journal editors and reviewers in the publication process.

The Afterword reconsiders the role of scientific publishing in the current, globalized world. Publishing a research paper brings a kind of social capital to the scientist, and science itself is recognized as knowledge capital. Science is acknowledged as a means to making the world a better place.

In sum, this book integrates the most current research concerning the role and production of scientific articles in English today. A global perspective and international focus means that it will be of interest to scientists and graduate students around the world.

Part I
The Context

Scientific research papers are among the most prestigious documents produced today. They are the means by which scientists report their scientific contributions. In addition, the publication of research papers is a means for scientists to establish their credibility. Each paper is counted as a product of knowledge. The number of knowledge products count toward determining a scientist's status, an institution's prestige and a nation's economic well-being. In other words, in the contemporary, globalized world of science, research papers have a value beyond their scientific content.

In this part, the context for writing and publishing research articles will be examined. The contexts are identified in four chapters:

- The rise of English as the language of science;
- Measuring the impact of articles, journals and nations;
- English competence, funds for research, and publishing success; and,
- Collaborations, teams and networks.

To explicate the context, we draw on empirical research in the fields of applied linguistics, bibliometrics, sociology of science, and economics. Together, they set the foundation for understanding the backdrop of contemporary scientific article creation. At the conclusion of this part, you will have a broad understanding of how this context influences the creation of research papers.

Chapter 1
The Rise of English as the Language of Science

Abstract English serves as the current language of science, however, it was not always so. Western scientific developments dating from the sixteenth century, such as Copernicus's work, were written in Latin. As the Enlightenment era took hold, all the languages of Western Europe were used for publishing scientific discovery. English played only a small part. Not until the post-World War I period did English begin to dominate scientific communication. The American scientific leadership also came into prominence as so much of Europe's science infrastructure was destroyed during two world wars.

Keywords Historic perspective • U. S. dominance • History of science • English • Scientific publishing • Enlightenment • Lingua franca

The scientific journals that carry the most prestige—*Nature, Science, New England Journal of Medicine, Cell,* and *Chemical Reviews*—are published in English. In fact, it is widely stated that English *is* the language of science. Of course, it wasn't always so.

In past millennia there was more than one language that was used for the sciences: Sumerian, Greek, Arabic and Latin (Hamel 2007). The Scientific Revolution of the sixteenth to eighteenth centuries that underlies the present scientific era is often marked as beginning with Nicolaus Copernicus. Writing in Latin, he dispelled the idea that the universe circled the earth in his 1543 work of *De revolutionibus orbium coelestium* (*On the Revolutions of the Heavenly Spheres*).

The scientific revolution during the Age of Enlightenment saw great change. Developments began to occur at a remarkable pace in the fields of mathematics, chemistry, human physiology, and physics. Communicating findings became central to the scientific endeavor. A principal outlet was the *Philosophical Transactions* of the Royal Society of London, established in 1660, and which published largely but not exclusively in English (Kaplan 2001, p. 7).

K. Englander, *Writing and Publishing Science Research Papers in English*,
SpringerBriefs in Education, DOI: 10.1007/978-94-007-7714-9_1, © The Author(s) 2014

At the beginning of the twenteith century, French, German, Russian and English all were used to communicate science, and multilingual science conferences were the norm. German was the dominant language of publication from 1900 to 1920 and was central to medicine, biology and chemistry (Hamel 2007). After the First World War the role of the German language was diminished, in part because it was banned at international conferences (Ammon 2001).

The prominence of English began as the prominence of German waned, and it coincided with the dominance of scientific research in the United States. The American influence in science, and thus the role of English, too, rose to prominence for several reasons. In the postwar period, the American educational and scientific infrastructure was still intact, while Europe's infrastructure had been devastated by war (Kaplan 2001). Large numbers of European scientists immigrated to the United States, and the so-called Cold War of the late 1940s and 1950s motivated vast investments in scientific exploration (Ferguson 2007). Together these factors allowed for English—the language of the United States with its economic and cultural power—to take on increasing importance for the dissemination of scientific knowledge.

Applied linguists rightly point out that English is not a *superior* language for science: it is not inherently any better than would be Latin or German or French, or Chinese (de Swaan 2001). English is not better suited to science because of innate qualities of the language. Rather, the confluence of socio-historic and economic factors gives English its contemporary dominance. There is no other language that is as commonly used in scientific forums, whether conferences, proceedings, research article publications, or citations.

It is interesting to note that some speculate that as China takes on increasing importance in global science, we may see Chinese become the language of science sometime toward the end of this century.

References

Ammon, U. (2001). Editor's preface. In U. Ammon (Ed.), *The dominance of English as a language of science* (pp. 5–10). Berlin: Mouton de Gruyter.

de Swaan, A. (2001). English in the social sciences. In U. Ammon (Ed.), *The dominance of english as a language of science* (pp. 71–83). Berlin: Mouton de Gruyter.

Ferguson, G. (2007). The global spread of English, scientific communication and ESP: Questions of equity, access and domain loss. *Ibérica, 13*, 7–38.

Hamel, R. E. (2007). The dominance of english in the international scientific periodical literature. *AILA Review, 20*, 53–71.

Kaplan, R. B. (2001). English—The accidental language of science? In U. Ammon (Ed.), *The dominance of English as a language of science: Effects on other languages and language communities* (pp. 3–26). Berlin: Mouton de Gruyter.

Chapter 2
Measuring the Impact of Articles, Journals and Nations

Abstract The rise of metrics that measure scientific production and impact is increasingly important. The system of determining 'impact' of a journal was created 50 years ago and the now ubiquitous 'impact factor' dominates much scientific publishing. The impact factor is used for judging journal importance although criticisms of this metric abound. Indices that count the number of scientific articles produced around the world indicate that the U.S. is no longer in first position, outplaced by Europe, and China has growing visibility in science article production.

Keywords Impact factor • Bibliometrics • Ranking • Web of Science • Metrics • Global standing • Rating scientists • Status • Google scholar • ISI • Citations • International journals

"Impact" is defined by the *Oxford English Dictionary* as "a marked effect or influence", and most scientists would hope that their work will have a marked effect or influence on their field. In science, having "impact" has become quantifiable through the "impact factor". The system of determining the impact of a journal was created in 1961 by Eugene Garfield, founder of the Institute for Scientific Information (ISI). The ISI is now incorporated into Thomson-Reuters' Web of Knowledge which includes the hugely important Science Citation Index (SCI) and Social Science Citation Index (SSCI). The impact factor, which Thomson-Reuters calculates, has become ubiquitous.

The formula for calculating a journal's impact factor is simple: (See Box 2.1)

Box 2.1 Calculation of Impact Factor

Number of citations to articles in a journal in one year (e.g. 2012)	÷	Number of articles published in that journal in the previous 2 years (e.g. 2010 and 2011)	=	IMPACT FACTOR

The reasoning behind this formula is simple: a journal publishes articles that are cited by others; therefore, the journal has a measurable impact on the field (Garfield 2006). The higher the impact factor, the greater the impact. Impact factors are recalculated every year.

K. Englander, *Writing and Publishing Science Research Papers in English*, SpringerBriefs in Education, DOI: 10.1007/978-94-007-7714-9_2,

Criticisms of the impact factor are numerous (See Box 2.1). Fundamentally, some argue that "citations are a shallow measure of research quality or impact" (Lillis and Curry 2010, p. 15).

Box 2.2 Criticisms of the Impact Factor

- A recent paper argues that the current metrics "undermine, rather than foster and reward, scholarship that matters" (Adler and Harzing 2009, p. 3).
- The two-year basis for counting citations may disadvantage journals or disciplines with longer publishing timelines, so that an accurate reflection of the number of citations an article would accrue is not seen within two years.
- Journals that publish many more articles, and disciplines that have many more journals, can obtain higher impact factors. Consequently, big journals in big disciplines are advantaged over smaller and highly specialized journals, and this may not be a true reflection of the importance of a journal.
- Young researchers who are trying to build a tenure and promotion file may avoid the more specialized journals with lower impact factors even though the work "would be better appreciated, published more quickly, and perhaps have more impact if they were published in specialized journals… This [practice] ultimately slows the diffusion of ideas into the research literature and stifles academic dialogue" (Segalla 2008, cited in Adler and Harzing 2009, p. 75).
- Journal policies sometimes encourage authors to cite other articles published by that same journal which can interfere with an objective indication of impact.
- Impact factors are used for purposes that were not intended. For example, impact factors are sometimes used for evaluating individuals (for the purpose of hiring, tenure and grant entitlement) and academic departments and institutions.

Nonetheless, the impact factor is now well entrenched in the world of scientific publishing. The impact factor is stated on the individual journal webpages of the world's four major journal publishers (Elsevier, Springer, Taylor and Francis and Wiley-Blackwell have well over 1000 journals each) (Ware and Mabe 2009). The impact factor is calculated for the 16,000+ journals included in the Web of Knowledge, comprised of the Science Citation Index Expanded, Social Sciences Citation Index Expanded, Conference Proceedings Citation Index, and Arts and

Table 2.1 Share of world articles and world citations by country/region

Country/region	Share of world articles (%)		Share of world citations (%)	
	1998	2008	1998	2008
World	100.0	100.0	100.0	100.0
United States	34.0	28.9	46.9	38.3
European Union-27	34.6	33.1	32.4	33.2
Japan	8.5	7.8	6.8	6.3
Asia-8	3.6	6.8	1.5	4.6
China	1.6	5.9	1.5	4.6

Humanities Citation Index (Thomson Reuters 2011). The field of bibliometrics, which has grown up around measuring impact, is central to the ranking of journals. For scientists to achieve maximum impact in the bibliometric system that is prevalent today, they are best to publish in the journals that are indexed in Thomson Reuters' Web of Knowledge.

A journal's impact factor is dependent upon being included in a prestigious citation index. The Web of Knowledge indexes 16,183 journals—a large number. Yet those journals constitute only 24 % of the "academic/scholarly" journals published in all languages that are included in Ulrich's Periodicals Directory (Lillis and Curry 2010, p. 17). Thus three-quarters of the scholarly journals published around the world are not counted in key international rankings of institutions and nations.

Certain types of journals are more likely to be included in the Web of Knowledge indexes. The index is "heavily biased" toward journals published in English from English-speaking countries (Lillis and Curry 2010, p. 18). Only 11.6 % of the journals that are included in the Web of Knowledge publish in a language that is not English (Brunner-Ried and Salazar-Muñiz 2012). The Web of Knowledge includes more minor U.S. journals than minor European journals, and relatively few non-English journals are included. In other words, a vast amount of scholarly publishing does not get included and therefore does not get counted.

Nonetheless the metrics determined by Web of Knowledge are used as the most authoritative and often the sole indicator of science knowledge production. Organizations that rank research publication output, such as the Organization for Economic Cooperation and Development (OECD) and the World Bank, rely almost exclusively on the journal, research article and citation data produced by the Thomson Reuters sources.

The statistics of production and attendant rankings reveal trends in world science, particularly when we take into account the limited number of journals counted in the Web of Knowledge. As shown in Table 2.1, the United States is now surpassed by the European Union in the share of world articles produced. However, papers published by American scientists continue to be cited in greater numbers than those by Europeans. Although China still ranks below Japan and the Asia-8, which includes India, it has produced a four-fold growth of papers to over 110,000 in the decade 1998–2008 (Pérez-Llantada 2010, p. 26).

Table 2.2 Publications and citations of 36 Latin American scholars reported in Web of Science and Google Scholar, selected from data in Brunner-Ried and Salazar-Muñiz 2012

Countries	Web of Science		Google Scholar	
	Number of journals	Number of citations	Number of journals	Number of citations
Argentina	9	29	715	6,202
Brazil	14	41	2,197	35,013
Spain	4	4	638	11,079
Mexico	11	43	1,264	12,796
United States	50	397	730	11,525

The research captured in the international indexes is also used to rank universities worldwide. Highly visible listings created by the *Times Higher Education* journal and the Shanghai Jiao Tong University determine university rankings by a mix of criteria, including the amount of research funds obtained, professor-student ratios, and number of degrees conferred. Significant weight in the determination is placed on "quality of staff" (i.e. number of highly cited researchers in a discipline) and research output as counted in publications and citations.

Citation counts are valuable in determining the importance of published work; these counts are used as a measure of the "quality" of the work. The importance of counting citations is based on the premise that the more frequently subsequent scholars refer to a paper, the higher the quality of that paper is. Much as the Web of Knowledge disproportionately favors journals published in English, also the citation counts that Web of Knowledge generates favor English-language citations.

The citation count differs when the Google Scholar citation search engine is used rather than Web of Knowledge. Google Scholar is used in many parts of the world as a search engine since access is free with internet connection. For example, the impact of 36 well-established Latin American scholars who have each been publishing for more than 30 years was compared in the two databases, Google Scholar and Web of Science (Table 2.2) (Brunner-Ried and Salazar-Muñiz 2012). A group of Argentinian academics, for example, received 29 citations in Web of Science and 6202 citations in Google Scholar. The immense difference can be due to the large number of Spanish-language journals included in Google Scholar and the fact that there are few in the Web of Knowledge. The large citation counts captured in Google Scholar seem to indicate that the impact of Latin American academics throughout the region is significantly more substantial than the Web of Science metrics indicate. In short, the picture of the impact of a scholar's work is altered when different citation searches are conducted.

The emphasis on publication metrics has created new demands and incentives for scientists in many parts of the world (Qiu 2010; Englander and Uzuner-Smith 2013). In China, scientists are awarded cash prizes, housing benefits or other perks for their publications in high profile journals. Failure to publish at least one paper per year can cause practicing doctors at a major surgical hospital in China to lose their medical privileges (Yongyan Li, personal communication, March, 2013). Systems of merit pay in Mexico and Turkey are linked directly to publishing research papers (Galaz-Fontes and Gil-Antón 2013; Englander and Uzuner-Smith 2013).

The pressure to "rack up publications" seems to encourage dubious research practices such as plagiarism, fabrication and falsification of data (Qiu 2010, p. 142). There is a concomitant rise in the number of retractions of published work, although these retractions were papers that all passed the peer review process. There is now a growth industry of "predatory publishers" which charge large sums of money to publish an author's work, provide no peer review and little or no database indexing (Rollin Kent, personal communication, May 26, 2013). Scientists may seek out the journals published by such publishers, either unwittingly or deliberately to get their work published. One biochemist expressed concern that "counting the number of publications, rather than assessing the quality of research, becomes the norm of evaluation" (Qiu 2010, p. 143). Thus individuals, institutions and nations all emphasize producing a high number of publications.

In sum, metrics have become central in determining value within science today. The emphasis on metrics of the number of articles and the citations they accrue is highly visible in institutional, national and international rankings. The metrics that underlie the rankings are calculated using indexes such as those created by the Web of Knowledge. There, journals gain visibility, since they are the journals that are included in the databases and thus are the journals that are identified when a scientist conducts a search for papers. The journals that are included provide the papers that are more likely to be consulted. Subsequently, those papers are more likely to be cited, raising the likelihood of obtaining or maintaining a high impact factor for the journal. The metrics very heavily favor publications in English, and publishing in English is more likely to produce citations in subsequent English-language articles. The desire on the part of nations, institutions and scientists to rank highly in the measures of research output and impact can give scientists much reason to be cognizant of the impact factors of the journals in which they seek to publish (Englander and Uzuner-Smith 2013).

References

Adler, N. J., & Harzing, A. W. (2009). When knowledge wins: Transcending the sense and non-sense of academic rankings. *Academy of Management Learning and Education, 8*(1), 72–95.

Brunner-Ried, J. J., & Salazar-Muñiz, F. (2012). Investigación educacional en Iberamérica: Entre la invisibilidad y la medición. *Magis Revista Internacional de Investigación en Educación, 4*(9), 559–575.

Englander, K., & Uzuner-Smith, S. (2013). The role of policy in constructing the peripheral scientist in the era of globalization. *Language Policy, 12*(3), 231–250.

Galaz-Fontes, J. F., Gil-Antón, M. (2013). The impact of merit-pay systems on the work and attitudes of Mexican academics. *Higher Education*, 8. Retrieved February 5, 2013.

Garfield, E. (2006). The history and meaning of the journal impact factor. *Journal of the American Medical Association [JAMA], 295*(1), 90–93.

Lillis, T., & Curry, M. J. (2010). *Academic writing in a global context: The politics and practices of publishing in English*. London: Routledge.

Pérez-Llantada, C. (2012). *Scientific discourse and the rhetoric of globalization*. London: Continuum.

Qiu, J. (2010). Publish or perish in China. *Nature, 463*, 142–143.

Thomson Reuters. (2011). http://thomsonreuters.com/products/ip-science/04_062/ssr1103443.pdf.

Ware, M., & Mabe, W. (2009). *The STM Report: An overview of scientific and scholarly journal publishing*. Oxford: International Association of Scientific, Technical and Medical Publishers.

Chapter 3
English Competence, Funds for Research and Publishing Success

Abstract Two dominant factors affect a nation's ranking in the production of scientific research papers: the funds provided for research and the degree of English-language competence. These factors provide great general advantage to scientists in the U.S. However, certain nations develop disciplinary expertise, such as Russia in physical sciences and France in mathematics. High TOEFL scores of non-English-speaking countries are a significant indicator of publishing success, outweighing research and development in some cases. Nations with low science capacity may seek cooperation with high science nations in order to conduct the research that can break cycles of poverty.

Keywords English • World rankings • R&D • GDP • Research-intensive universities • Tier 1 • Citations • ESL • TOEFL

Chapter 2 demonstrated the importance of publishing in journals of high impact factor. We discussed the ranking of scientists, institutions, and nations that follows from publishing in journals that are included in indexes such as the Web of Knowledge Science Citation Index. Here, two other elements that contribute to the likelihood of publishing success in such journals are discussed: funds for research and English competence.

The number of published research papers is closely correlated with public expenditure on research and development. By far the greatest number of dollars and highest percentage of Gross Domestic Product (GDP) is invested in research in the U.S. In fact, the U.S. contributes 41 % of the global funds that are invested in research; next is Japan at 15 %, China with 12 %, and Germany with 8 % (OECD 2011).

Funds alone do not account for the strong performance of the United States in research, however. The post-World War II period with its stimulus of Cold War-motivated research and the "space race" supported the creation and spectacular

K. Englander, *Writing and Publishing Science Research Papers in English*, SpringerBriefs in Education, DOI: 10.1007/978-94-007-7714-9_3,

growth of a research and development infrastructure. Universities were notable beneficiaries of this investment.

Now, some 60 years later, the strength of American universities is dominant. Measures that rank the top 50 universities worldwide, i.e., those that have the highest impact, place 40 of those universities in the United States (OECD 2011). The other 10 are in Europe. Within the United States, two dozen universities account for almost half of the entire nation's research output (Mervis 2010). Research is not only dominant in the U.S., the majority of the research is concentrated in only a few universities. Although the United States is the strongest nation in research performance, the European Union's aggregate output of research papers now surpasses that of the U.S. There is also evidence that some universities in Asia are emerging as leading research institutions (OECD 2011).

Citation counts are indicators of the impact of published research. Interestingly, sheer size of article output does not account for citation counts, as seen in the statistics for "wealth intensity" in relation to "citation intensity" (King 2004). Wealth intensity is the GDP per person. When this is measured against citations, several European nations outrank the U.S. (Scandinavia, the Netherlands and Switzerland), as does Israel. This indicates that despite the smaller investment in research, some European nations produce research of great impact.

Globally, stark regional variation exists. The seven countries with the highest citation counts (U.S., U.K., China, Germany, France, Canada, and Japan) account for the overwhelming proportion of the world's citations. There is poor representation of the rest of the world, and this can be seen in regional data. The African country that reaches the highest is South Africa which ranks at number 35, followed by Kenya at number 54. Brazil is the highest ranking South American country, at number 17. Of the Islamic countries, Iran is number 23, despite such prominent Nobel prizewinners as Abdus Salam (physics, 1979) and Ahmed Zewail (chemistry, 1999).

Some nations perform strongly in particular disciplines. King (2004) reports that:

- Russia is relatively strong in the physical sciences and weak in life sciences;
- Japan is strong in physical sciences and engineering;
- France is strong in mathematics;
- Germany has highest impact in the physical sciences;
- The United Kingdom has highest impact in medical, life and environmental sciences, high in mathematics, but not strong in physical sciences and engineering;
- The U.S. publishes disproportionately in health and social sciences (Mervis 2010).

There is another factor beyond investment in research that has a significant effect on publications in non-English-speaking nations: English language competence. Man et al's (2004) study examined the top five medical journals and the geographic location of the authors of the published articles over a five-year period (1996–2001). While total research funding as a percentage of GDP was significantly correlated with publication rate (correlational coefficient of 0.44, $p = 0.05$), the effect of English was even larger. English competence was determined based

Table 3.1 TOEFL rank and research spending of non-English-speaking OECD nations (Man et al. 2004)

Countries	TOEFL rank	Research spending rank	Publication rank
The Netherlands	1	8	4
Denmark	2	9	1
Belgium	3	10	6
Austria	4	14	8
Germany	5	6	10
Finland	6	3	2
Sweden	7	1	5

on national TOEFL (Test of English as a Foreign Language) scores. High TOEFL scores were correlated with publication rate even more significantly (correlational coefficient of 0.77, $p = 0.01$) than research funding. Even among the privileged nations of the OECD, publication rates were more greatly correlated with English competence than research spending (Table 3.1).

The world's English-speaking nations and the nations of Western Europe are distinctly greater producers of scientific research articles than all other parts of the world, with the exception of China and Japan, as noted above. This disparity has immense implications for science and development. A disproportionate concentration of science is conducted in the developed world. Yet the developing world has tremendous local and national needs in healthcare, clean water, sanitation, food and transportation require capabilities in engineering, technology, medicine, and social sciences. The combination of few science resources, a small number of journals included in the international databases, and little recognition in the form of citations, assures that developing nations have limited ability to address national needs. They seem to also have limited access to the international networks that provide exposure for scientific accomplishment. Consequently, nations of low science capacity require "capacity building" in cooperation with high science nations to do the research that can break cycles of poverty (King 2004, p. 315).

Developing teams, particularly international ones, can be one means of providing the scientific exchange that permits national development and scientific recognition. The importance of teams, international collaboration and participating in networks is discussed in the next chapter.

References

King, D. A. (2004). The scientific impact of nations. *Nature, 430*, 311–326.

Man, J. P., Weinkauf, J., Tsang, M., & Sin, D. (2004). Why do some countries publish more than others? *European Journal of Epidemiology, 19*(8), 811–817.

Mervis, J. (2010). Handful of U.S. schools claim larger share of output. *Science, 330*, 1032.

OECD (2011). *OECD Science, Technology and Industry Scoreboard.* Paris: Organization of Economic Cooperation and Development.

Chapter 4
Collaborations, Teams and Networks

Abstract Collaborations in research and writing-up that research are now the norm in science. The number of papers authored by teams and number of authors per paper is increasing. Team-authored papers are cited more frequently, indicating greater influence in the discipline. Teams provide advantage to the lower-ranked institution in cross-border and cross-ranking collaborations. Formal team relationships differ from informal network relationships. Informal networks often prove successful for scientists outside Anglophone countries when they seek to publish their work.

Keywords Collaboration • Multiple authors • Research networks • Informal networks • Single authored papers • Big science • Research team

Science as an endeavor is not a solo activity. Despite the image in popular culture of the lone scientist enveloped in a white lab coat working endlessly with tubes, beakers and instrument panels who then arrives at the eureka moment, rarely is science performed in this way. In this chapter, the evidence concerning scientific collaboration, international teams and disciplinary networks will be discussed. In short, better connections lead to greater prominence and, often, better science.

There is substantial evidence that working in collaboration with others is the dominant manner of work for most scientists, and this trend is increasing. In a study of 19.9 million research articles over five decades, the growth of multiple-authored papers is clear (Wuchty et al. 2007). In the sciences, in 1955, average team size was 1.9 authors; in 2000, team size doubled to 3.5 authors per paper. Papers with a great number of authors are increasing as well (Pérez-Llantada 2012) (See Table 4.1).

Both the number of papers authored by teams and the number of authors per paper is steadily increasing. Further, the trend is seen not only in laboratory sciences where teamwork might be expected, but also in mathematics, which is often thought to be a solitary discipline. In other words, teamwork is growing in every

K. Englander, *Writing and Publishing Science Research Papers in English*,
SpringerBriefs in Education, DOI: 10.1007/978-94-007-7714-9_4, © The Author(s) 2014

Table 4.1 Research paper growth with large numbers of authors (based on Pérez-Llantada 2012)

Number of authors	2003/4	2005
50 or more	500	750
100 or more	300	475
500 or more	40	131

subdiscipline included in the Web of Science. Co-authorship is increasing in the social sciences (economics, sociology) and the humanities field of philosophy, as well, but to a lesser degree (McNamee and Willis 1994).

The question arises as to whether team-authored papers lessen the impact of the work itself, because as the writer F. Scott Fitzgerald has written, "No grand idea was ever born in a conference" (cited in Wuchty et al. 2007, p. 1036). Citation patterns indicate that team-authored papers are actually more influential, particularly in the twenty-first century. Team-authored papers receive more than twice the citations that a single-authored paper has, even when correcting for self-citation.

Highly cited work (i.e. more than 1000 citations) is another indicator of impact (Wuchty et al. 2007). In the 1950s, single authored papers were more likely to be extraordinarily cited, but the situation is different now. In 2000, a team-authored paper was six times more likely to be highly cited than a solo paper. We see that teamwork is now the dominant structure for doing science, and for doing good science.

Collaboration in the United States very often crosses institutional boundaries (Jones et al. 2008). Collaborations in science and engineering between tier I schools (Table 4.2) are more likely to be of high impact, as measured by citation counts, than a paper produced solely within one tier I school (6.9 % more likely, $p < 0.0001$). Collaborations between a tier I school and tier II school (i.e. top 6–10 % of universities) tend to be of lesser impact, but still greater than collaboration among two tier II schools. The impact of collaboration across tiers (for example, between a tier I and tier III) tends to measure closer to the higher tiered school rather than the lower. Thus it seems advantageous to collaborate 'up' where possible. It should be remembered that even tier III schools constitute the best the U.S. has to offer to the global research world. Elite universities which collaborate with each other form a nucleus of science research and production of immense impact. The evidence indicates that "social distance" is of increasing importance in research collaborations (Jones et al. 2008, p. 1261).

Table 4.2 Tiers of U.S. universities (based on Jones et al. 2008)

Tier	Ranking as percentage of U.S. universities
I	Top 5 % of U.S.
II	Top 6–10 %
III	Top 11–20 %
IV	Remainder

Collaboration across national boundaries is generally increasing, too (Pérez-Llantada 2012). Based on articles which were published in journals in the Web of Science, there was a 10 % increase in U.S. collaboration with authors outside the U.S. in the period 1998–2008. France's international collaborations increased 14 %; Spain's 11 %. Interestingly, China decreased 1 % in that decade, but other Asian countries also increased.

The tremendous amount of equipment and resources necessary to conduct "big science" and the interdisciplinary and multidisciplinary teams often required seem to be fueling the large team sizes and collaborative activities (Galison and Helvy 1992). When the collaboration is an international one, advantages can accrue to both nations in international collaborations, but the advantages may differ (Hwang 2008). When there is an imbalance of status between the collaborating countries, there may be advantages of prestige and technical expertise garnered by the less advanced country. In exchange, the more advanced country may be able to work with the best and the brightest researchers in the host country and apply established models to local contexts.

In addition to collaboration, being part of networks of scientists has been shown to be important in getting work seen and published (Lillis and Curry 2010). The networks which a scientist is part of can be local or international. They can be weak (informal relationships) or strong (formal research relationships). The information that is exchanged in such networks can include word-of-mouth calls for submission to journal special issues, participation in conference symposia, inside knowledge on funding opportunities, and even new job opportunities. These conversations seem quite commonplace among insiders. For novice or international scientists, the conversations are a crucial opportunity to gain insider knowledge: knowledge that can lead to publication and other activities within the discipline. The conversations have been described as "social practices that surround scholarly texts" (Canagarajah, in Englander 2011, p. 3). Participating in these social practices are highly valuable for enhancing publishing and creating a presence in the discipline.

In summary, in this part of the book, the context of publishing scientific articles has been examined. The context is foremost an era in which English has a dominant role as the language for publishing science, although this was not always the case. This is an era of knowledge economies, and producing knowledge improves a nation's position in the global arena. Knowledge production is most commonly quantified as research papers that are published in the journals that are included in a few highly prestigious databases, such as the Web of Knowledge. Journals are identified by impact factors, which are used to rank prestige of the journal, and the citations accrued are commonly used to rank scientists and their papers. Thus the current context for publishing research papers is highly determined by metrics. Analysis of high impact papers indicates the growing trend for team publishing and across-institution publishing. The highest impact papers tend to be authored as collaborations between researchers at two tier I schools in the U.S. There is growing trend towards collaboration across national boundaries, as well. The prominence of metrics in evaluating researchers' prestige and importance places increasing importance on developing networks within the discipline in order to take advantage of the knowledge that insiders have about the field and opportunities for fruitful collaboration and publishing. This context of collaboration, networks, and metrics has been increasing steadily in the last decade or two. In this same period, the United States has lost its dominance as the producer of the greatest number of published research papers. Europe and, soon, Asia are out-publishing the U.S. Scientists around the world are making decisions to publish their work in the journals that count in the international databases, and the vast majority of those journals are published in English. In Part II of this book, the research article itself is examined, focusing especially on English-language research articles.

References

Englander, K. (2011). The ideal situation: To be located in your own country and be a part of the conversation—an interview with Suresh Canagarajah. *MEXTESOL Journal, 35*(2), 1–5.

Galison, P., & Helvy, B. (1992). *Big science: The growth of large-scale research*. Stanford: Stanford University Press.

Hwang, K. (2008). International collaboration in multilayered center-periphery in the globalization of science and technology. *Science, Technology and Human Values, 23*, 101–133.

Jones, B. F., Wuchty, S., & Uzzi, B. (2008). Multi-university research teams: Shifting impact, geography and stratification in science. *Science, 322*, 1259–1262.

Lillis, T., & Curry, M. J. (2010). *Academic writing in a global context: The politics and practices of publishing in English*. London: Routledge.

McNamee, S. J., & Willis, C. (1994). Stratification in science: A comparison of publication patterns in four disciplines. *Science Communication, 15*, 396–416.

Pérez-Llantada, C. (2012). *Scientific discourse and the rhetoric of globalization*. London: Continuum.

Wuchty, S., Jones, B. F., Uzzi, B. (2007). The increasing dominance of teams in production of knowledge. *Science, 316*, 1036–1039.

Part II
The Text

Part II of this book examines the scientific paper as a document. Our review of research is based on studies conducted by applied linguists, socioanthropologists, rhetoricians, sociologists and philosophers interested in scientific research articles. All the research presented in the following five chapters is based on the premise that the scientific article is a carefully constructed document that plays an essential role in the creation and dissemination of scientific knowledge. Because the research paper is a constructed document, it can be analyzed in relation to the linguistic and discursive features that are typical of this literacy form. Our understanding of the scientific article and its relationship to the act of producing scientific knowledge mean that we assume that embedded within the scientific article are the underlying epistemological assumptions of science. These assumptions manifest themselves in the particular forms present in this writing act. We also presume that the document that is published in an English-language journal is the result of a negotiation that occurs between the writer–scientist and the journal editors and reviewers. Therefore, in explicating the scientific article in this brief review we will address the linguistic, discursive, epistemological and interactional aspects of this form of writing.

Chapter 5
The Scientific Research Article and the Creation of Science

Abstract A scientific research paper is more complex than a simple reporting of events. It is a document written for an expert audience that represents the science community's notion of valid argumentation. In this sense, it is a construction intended to persuade the reader of the validity of the claims. A historical review of scientific papers shows that the form of papers has changed dramatically. The current standardized structure allows the knowledgeable reader to access information quickly, identify the paper's strengths and weaknesses and use it as a basis for further scientific work.

Keywords Rhetoric • Style • Argumentation • History of science articles • Reporting science • Royal Society • German

Science writers often claim that the science paper is a record of what was done in the lab. They claim that the paper is an accurate rendering of the hypothetico-inductive process that scientists go through in their research. This process is comprised of a series of discrete, successive steps in which the scientist makes a hypothesis, performs an experiment, takes measurements, interprets the results, formulates models and theories, and thereby reaches conclusions. This process is then faithfully reported in a research paper in order to communicate to the larger scientific community the new scientific knowledge.

Science is rarely performed that way, and the scientific paper is not a simple reporting of what occurred (See Box 5.1). Two respected sociologists who have studied laboratory practice say that "actual scientific practice entails the confrontation and negotiation of utter confusion" (Latour and Woolgar 1986, p. 36). In extreme form, the discrepancy between what occurred in the lab, and the published scientific paper has been termed a "mismatch" (Schickore 2008, p. 323) where the final paper is a rational reconstruction that creates the appearance of a chronology of scientific activity (Berkenkotter and Huckin 1995; Schickore 2008).

K. Englander, *Writing and Publishing Science Research Papers in English*,
SpringerBriefs in Education, DOI: 10.1007/978-94-007-7714-9_5, © The Author(s) 2014

Box 5.1 Criticisms of Scientific Articles as Representations of the Scientific Process

For practicing scientists, the view of philosophers (Reichenbach 1938) historians and social anthropologists (Knorr-Cetina 1981; Latour and Woolgar 1986) about science may appear exaggerated, or even uninformed. Nonetheless, we summarize some key ideas about the "mismatch" between doing science and reporting of science to highlight the fact that research papers are constructed.

Some claim that scientific papers provide no reliable information about the nature of the practice of science and therefore science educators should not use them as the basis for educating scientists about how to perform science.

The scientific paper is "a fraud in the sense that it does give a totally misleading narrative of the processes of thought that go into scientific discoveries" (Medawar 1966, p. 38). Scientific papers do not recount the events in a true-to-life manner: first we did this, then we did this, then we found this and so we changed and did that. Because of the discrepancy between what "really" happens and the reconstruction in science papers, some say that the current format for reporting science should be discarded (Medawar 1966, p. 38).

The research paper should document the *actual* scientific process, including the hypotheses that were eliminated, the methods which did not work (Egger et al. 1997) and the true motivation for a particular study, such as the fact that the lab purchased a new piece of equipment and so it was important to devise an experiment to take advantage of it (Knorr-Cetina 1981).

While mismatch may be an inflammatory description, it would be inaccurate to claim that a research paper is a simple reporting of events. There is a difference between scientific activities and the acceptable forms of scientific publishing. Doing science and writing science are related, but distinct. Scientists, and the social science researchers who study them, seem to agree on four important points concerning research papers (Fig. 5.1, based on Schickore 2008).

There is agreement among scholars regarding the importance, role, readership and argumentation of research papers. In other words, a scientific paper is a text that has evolved to fulfill communicative needs in a social setting, and it is constructed with a purpose (Hanauer 2006). The scientific research paper is considered a "genre" because it has a structural form that meets the dynamic needs of a community of practitioners and in doing so it fulfills social functions (Miller 1984). Broadly speaking, the purpose of the research paper is to disseminate scientific knowledge. It is constructed in the manner that is currently expected by scientific peers, although there is variation among science disciplines and specific journals.

The structure, style, and content of scientific papers have not been consistent over time. As we documented in Chap. 2, English has not always been the

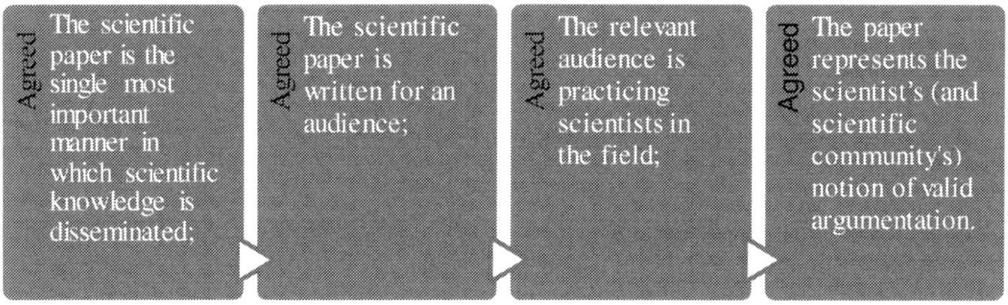

Fig. 5.1 Four points of agreement concerning scientific research papers, based on Schickore (2008)

language of science. The research paper has not remained the same over the past 300 years either. For example, the readers of papers in the Royal Society of the French Academy of Sciences would see exhaustive details of methods because little consensus had yet been reached about what constituted a proper methodology. Personal testimony was typical in early papers (Atkinson 1999; Gross et al. 2002). German researchers in the 1800s were based in far-flung universities and so developed a style of argumentation quite distinct from today's presentation of evidence (Schickore 2008).

The genre of the scientific paper today continues to evolve. Varied kinds of journals impose different writing styles on the scientists. Three simple examples are journals that expect an 8,000-word paper, versus *Letters* that require papers to be very short, versus electronic journals that allow papers to be hypertexted with extensive supplementary material. The ubiquity of the internet has seen the development of journals that ask scientists to post videos of the Method section of paper (Lunsford 2011). Thus, writing a scientific research paper occurs in a specific context, in a particular era, for an audience with different degrees of shared knowledge and scientific sophistication.

In summary, the scientific research paper is not a transparent recounting of what was done that allowed the scientist to reach the conclusions that are reported in the paper. Social anthropologist Karin Knorr-Cetina points out that the scientific paper "forgets much of what happened in the laboratory while purporting to be a 'report' of that research" (Knorr-Cetina 1981, p. 94). Instead, the research paper is a carefully constructed document that complies with the demands and expectations of the genre. It presents results based on scientific work that was undertaken in a sequence quite different from the inductive linear one of the narrative, and it serves to convince or win over peers to the knowledge claim which is argued. As such, the scientific article is an accurate, constructed presentation of what was done in the research process and the results that emerged from that process. It has a standardized form that is accepted within the discipline. Because of this standardization, although it may change over time, the knowledgeable reader can access the information within it quickly, identify the research strengths and weaknesses, and ultimately use it as a basis for further scientific work.

References

Atkinson, D. (1999). *Scientific discourse in a sociohistorical context: The Philosophical Transactions of the Royal Society of London, 1675–1975*. Hillsdale: Erlbaum.

Berkenkotter, C., & Huckin, T. (1995). *Genre knowledge in disciplinary communication: Cognition, culture, power*. Mahwah: Erlbaum.

Egger, M., Zellweger-Zahner, T., Schneider, M., Junker, C., Lengeler, C., & Antes, G. (1997). Language bias in randomised controlled trials published in English and German. *The Lancet, 350*, 326–329.

Gross, A. G., Harmon, J. E., & Reidy, M. S. (2002). *Communicating science: The scientific article from the 17th century to the present*. West Lafayette: Parlor Press.

Hanauer, D. I. (2006). *Scientific discourse: Multiliteracy in the classroom*. London: Continuum.

Knorr-Cetina, K. (1981). *The manufacture of knowledge: An essay on the constructivist and contextual nature of science*. Oxford: Pergamon.

Latour, B., & Woolgar, S. (1986). *Laboratory life: The social construction of scientific facts*. Princeton: Princeton University Press.

Lunsford, K. (2011). Professional development of postdoctoral scientists. Writing research across borders II. Fairfax, VA.

Medawar, P. B. (1966). Is the scientific paper a fraud? In P. B. Medawar (Ed.), *The strange case of the spotted mice and other classic essays on science* (pp. 33–39). Oxford: Oxford University Press.

Miller, C. R. (1984). Genre as social action. *Quarterly Journal of Speech, 70*, 151–167.

Reichenbach, H. (1938). *Experience and prediction: An analysis of the foundations and structure of knowledge*. Chicago: University of Chicago Press.

Schickore, J. (2008). Doing science, writing science. *Philosophy of Science, 75*(3), 323–343.

Chapter 6
Varieties of Science Texts

Abstract Scientific research papers are only one of many different kinds of documents that scientists write. The interrelated documents form a chain or network that can precede, accompany and follow the research paper. Some documents are public, such as the research paper or the conference presentation. Other documents are occluded, in that they are seen by few people, such as grant applications or manuscript referee comments. By recognizing the genre chain, a scientist can plan the content, audience and timeline of the various texts that are related to a scientific investigation.

Keywords Conference abstracts • Grant proposals • Reviewers • Drafts • Genre • Occluded texts

The scientific research paper is not the only kind of document that scientists need to write. Rather, there is a chain or network of interrelated documents that precede, accompany and follow the research paper. This interrelationship of documents, called a genre chain (Fairclough 2003; Swales 2004), takes elements of the science investigation process and rearranges them for different purposes and different audiences. For example, the research grant proposal is written for rigorous evaluators about work that has not yet been performed and that requires money over a specified period of time. This is a high stakes genre that can affect the researcher's work for quite a long time. The conference proposal/abstract has much lower stakes for the researcher. The conference proposal is also written for evaluators, but there isn't a request for money and the consequences of rejection are typically lesser. It is a precursor to the genre of the oral presentation genre that is usually accompanied by some sort of visual support such as PowerPoint. A sample genre chain is shown in Fig. 6.1.

The research grant proposal and the conference proposal are not publicly available documents. They are occluded (Swales 2004) in the sense that only the people immediately involved see the documents, such as the writer and evaluators. The occluded

K. Englander, *Writing and Publishing Science Research Papers in English*, SpringerBriefs in Education, DOI: 10.1007/978-94-007-7714-9_6, © The Author(s) 2014

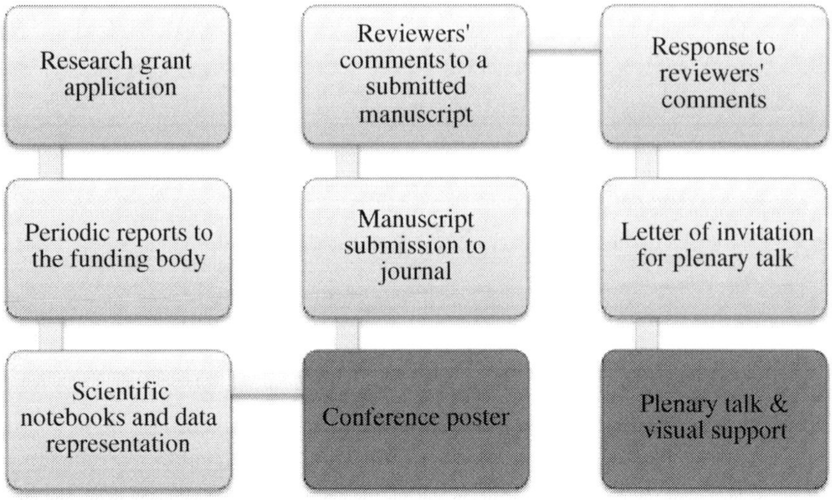

Fig. 6.1 Sample genre chain composed of occluded documents (*light gray boxes*) and public documents (*dark gray boxes*)

documents are behind-the-scenes and they surround the public ones. For example, the occluded submission letter and reviewer comments surround the published research paper. The occluded conference abstract connects to the public conference presentation and accompanying PowerPoint slides. The fact that occluded documents are not easily accessible can make it difficult to properly anticipate and produce them, especially for junior researchers and graduate students.

Knowing the interrelationship between the documents assists the scientist in planning time to produce them and respond to them. The depiction in Fig. 6.1 is a simplified chain. Many other documents may enter into the chain–press releases, letters to *Nature,* purchase requisitions, reference letters of postdocs. The number and range of occluded genres in science is noticeably greater than the public genres.

The value of recognizing the genre chain is to be able to identify the range of oral and written documents that a scientific investigation affects. Applying knowledge about who needs to know what as the recipients of the documents allows the scientist to adjust the content, logic and style when producing those documents. Clearly, the science research paper is only one of the interrelated genres in the chain, but it is a genre with a prominent public presence and which is used to establish the worth of the science itself.

References

Fairclough, N. (2003). *Analysing discourse: Textual analysis for social reserach.* New York: Routledge.

Swales, J. M. (2004). *Research genres: Explorations and applications.* Cambridge: Cambridge University Press.

Chapter 7
Structure of the Research Article in the Creation of Knowledge

Abstract Research papers are more than impartial presentations of objective knowledge. They possess a social significance that aims to persuade the reader of the validity of claims. Three linguistic devices are common in research papers to support the rhetorical aim. The appropriate level of claim is achieved by various strategies of modulation. Passive sentence structures focus attention on the scientific actions and processes rather than on the scientist. Complex noun structures permit a great deal of information to be communicated with few words, allowing the expert reader to grasp ideas quickly.

Keywords Making claims • Passive • Noun structures • Style • Expert reader • Nominalization • Structure of papers • Style • Rhetoric • Hedging • Boosting

The research paper is a written genre that disseminates knowledge to the broader scientific community. As such, it is of great social significance (Zerbe 2007). However, to describe the paper's role as simply "disseminating" knowledge does not do justice to the genre. Dissemination implies an impartial presentation of objective knowledge (Hanauer and Englander 2013). In fact, the content and style of writing in a research paper involve "a good deal of literary strategy largely unnoticed by the reader" (Knorr-Cetina 1981, p. 94). The research paper is constructed.

The literary strategy that underpins the scientific paper is characterized as a rhetoric of persuasion (Bazerman 1988). A scientific paper serves to influence the reader in such a way that that there is no alternative conceptualization of the world than the one presented in the text (Bazerman 1988). It works to persuade the reader that the claims should be considered to be factual. By establishing the claims as factual, they can then be cited in subsequent papers by the researchers themselves and, importantly, by others.

To explore how particular structures of the research paper genre serve to create knowledge, three linguistic aspects are discussed here. Those structures are

K. Englander, *Writing and Publishing Science Research Papers in English*, SpringerBriefs in Education, DOI: 10.1007/978-94-007-7714-9_7, © The Author(s) 2014

making claims, personalization and passives, and noun structures. In the next chapter, the sociolinguistic features of each section of the typical Introduction-Methods-Results-Discussion format are examined.

7.1 Making Claims

Making an acceptable claim is a complex activity that involves the members of the research team and the larger scientific community in the form of journal reviewers and editors. The role of journal reviewers and editors is discussed more thoroughly in Chap. 13, but here we want to point out the role they have in modulating knowledge claims.

Claims are characterized on a continuum from higher level to lower level (Pinch 1985). Higher level claims may be profound but also risky, whereas lower-level claims are likely correct but considered to be trivial. Depending the body of literature in which a paper is situated and the audience it reaches, the same claim can be considered by reviewers as 'speculative' or 'well-defined', a 'highly significant' advance or a 'well known' observation (Myers 1985, p. 596). This variation in interpretation can distinguish between writing for a narrowly-defined disciplinary audience or a broader multidisciplinary one. Attempting to publish higher level claims in high prestige journals that are multidisciplinary, such as *Science* or *Nature*, is often a negotiated process between the scientist and the editor or reviewers, and it can be elusive even for established scientists (Myers 1985).

A hierarchy of claims was identified by science ethnographers Latour and Woolgar (1986). There are five levels or types of statements that move from the most speculative to the most well established statement (Fig. 7.1).

The scientific enterprise, based on an analysis of laboratory work, is to move from Type 1 speculative statements to the definitive Type 5 statements. The scientist aims to produce papers with few Type 1, 2, and 3 statements in order to make higher level claims. The scientist's goal is to have one's own claims become the Type 4 statements that are cited by others (Latour and Woolgar 1986).

One means of adjusting levels of claim is by the degree of generalization or specificity in a statement. Myers (1985, p. 603) illustrates the different scope of possible claims by the zoologist Dr. Crews (See Fig. 7.2).

As shown in Fig. 7.2, as the scope of the generalizability is constrained, the claim becomes more specific. At the lowest level of generalization, there is great specificity to one species, the context, and one particular phenomenon. In the mid-level claim, three phenomena are named and a broader categorization of species is stated. In the greatest generalization, a large category of factors has a two-fold impact on three aspects of the phenomenon. The broader the scope, the greater the generalizability.

Based on these two Figs. 7.1 and 7.2, we see that there are two interrelated phenomena that determine the construction of the knowledge claim: (i) the degree of certainty accorded to the claim (from speculation to uncontested facts), and (ii) the breadth of generalization that the content affects. To have the widest

Type 5	• Statements are "taken-for-granted" to such an extent that experts assume them to be basic facts and do not merit discussion.
Type 4	• Statements are well established but they are explicitly stated. In introductory science textbooks, they are presented as factual statements.
Type 3	• Statements add modality (could, may, might) to facts and thereby calls into question the factual nature of a claim. These statements are often found in previous reviews of the literature.
Type 2	• Statements are more tentative and present reasonable conclusions based on the evidence.
Type 1	• Statements are speculative and consist of conjectures.

Fig. 7.1 Taxonomy of statements from most speculative to most definitive, based on Latour and Woolgar (1986)

Greatest degree of generalization	• "Environmental factors may influence the evolution and development of three aspects of reproduction: (i)..., (ii)..., (iii)...."
Lesser generalization	• "Environmental factors may cause gamete production, sex hormone secretion, and mating behavior to be dissociated."
More specificity	• "Gamete production, sex hormone secretion and mating behavior are dissociated in some species of each class of vertebrates."
Further specificity	• "Gamete production, sex hormone secretion and mating behavior are dissociated in the red -sided garter snake."
Lowest level of generalization	• "The red-sided garter snake mates at the beginning of warm weather, when sex hormone levels are low."

Fig. 7.2 Scope and levels of claims for a biologist, with examples, based on Myers (1985)

impact, scientists aim for the greatest certainty and the broadest generalization that their research and the journal editor will allow.

Claim statements as linguistic entities can be made stronger or weaker, and this is referred to by linguists as boosting (sometimes called 'emphatics') or hedging, respectively. Hedging is seen in the Type 3 statements above, in which the writer adds modals such as 'could' or 'may' or 'might' to a statement and thereby makes it less definitive. Hedging indicates that "a statement is based on plausible reasoning rather than certain knowledge" (Hyland 2005, p. 179). For example,

> Our results suggest that rapid freeze and thaw rates during artificial experiments in the laboratory **may** cause artifactual formation of embolism. Such experiments **may** not quantitatively represent the amount of embolism that is formed during winter freezing in nature. In the chaparral **at least**, low temperature episodes **usually** result in gradual freeze–thaw events (cited in Hyland 2005, p. 179, emphasis in original).

In the example above, other linguistic elements are also used to mitigate the certainty of the statements. The word "suggest" carries less certainty than saying "our results indicate" or "demonstrate". The choice "suggest"—a reporting verb—indicates the writer's certainty and stance toward the content. Other adverbs, too, are seen in the example above ("at least" and "usually") to constrain the generalization regarding where low temperature episodes occur and how frequently they affect the freeze–thaw events. Often hedges appear in clusters, creating a cumulative effect whereby the claims seem to be reasonable and considered.

To create emphasis and certainty, boosters are typically used. They also serve to show involvement or solidarity with the reader. Boosting is accomplished with adverbs such as "clearly" and "undoubtedly" and "obviously". For example,

> This brings us into conflict with Currie's account, for static images **surely** cannot trigger our capacity to recognize movement. If that were so, we would see the image as itself moving. With a few interesting exceptions we **obviously** do not see a static image as moving. Suppose, then, that we say that static images only depict instants. This too creates problems, for it suggests that we have a recognitional capacity for instants, and this seems **highly** dubious (cited in Hyland 2005, p. 179, emphasis in original).

Hedging and boosting can be accomplished with what Grabe and Kaplan (1997) categorized as verbs, non-verbs and emphatic expressions. Examples of the three categories taken from natural science journals are shown in Table 7.1.

The linguistic manifestations of hedging and boosting have been examined extensively by the applied linguist Ken Hyland (Hyland 2000a, b, 2005). His analysis of papers in biology, chemistry and physics indicate that hedges always outnumber boosters, but both always appear in the science papers. The most common hedges and boosters are shown in Table 7.2 (Hyland 2000b).

Table 7.1 Linguistic expressions of hedging and boosting, based on Grabe and Kaplan (1997, p. 162)

Verbs for hedging and boosting				
claim	assume	say	report	suggest
appear	seem	argue	find	believe
Non-verbs for hedging and boosting				
sometimes	most of	typically	nearly	in large part
likely that	possibility	approximately	evident that	
Emphatic expressions				
of course	virtually all	in fact	indeed	finally
clearly	surely			

Table 7.2 Hedges and boosters listed in decreasing order of frequency, based on Hyland (2000b, p. 91)

Hedges	Boosters
would	show (that)
may	find (that)
suggest	determine
indicate	will
about	it is clear/clearly
assume	the fact that
possible/possibly	demonstrate (that)
could	confirm

With this variety and range of linguistic possibilities for stating one's knowledge claims, the writer's options in formulating the claim are decidedly complex. The decision must take into account not only the level of generalization and level of certainty that the writer wishes to make, but it must also be acceptable to the journal editors and reviewers who determine the paper's acceptability for the journal's readers.

It is important to note that adjusting claims to the intended level can be especially difficult for non-native speakers of English. This can be the result of two related phenomena. One difficulty is that the conventions of what is the appropriate amount of hedging or boosting can be different in another language community than the conventions of English science writing. Studies that identify different frequencies of hedging and boosting have been conducted regarding quite a number of languages:

- In German, boosts are common for "assertion and authority" (Kreutz and Harres, 1997);
- In Bulgarian, hedges are used in different proportions than in English (Vassileva 1997);
- German and Finnish social science writers use more hedges than English writers (Crismore and Vande Kopple 1997);
- Sudanese (ElMalik and Nesi 2008) and Korean (Cho 2009) scholars avoid "boasting" and so they limit boosters;
- Mexican Spanish-speaking scientists are accused of being too hyperbolic in claims because they use fewer hedges (Englander 2009);
- Chinese scholars based in Hong Kong are sometimes considered "too assertive" with few hedges in their papers (Flowerdew 1999).

Because the conventions within the science writing community in a particular language are different, adjusting to the amount of hedging or boosting in English can be uncomfortable for the non-native-English writer (A more thorough examination of differences in published papers in different languages and cultures is presented in Chap. 9).

The other difficulty for non-native-English writers is knowing how to use English to convey the appropriate levels of caution, humility and importance regarding the work itself and the writer's relationship to the reader. In part, this uncertainty may occur because some studies show that when readers read, they do not give much attention to the way hedges and boosters affect the content (Hyland 2000a, b). Non-native-English speakers may focus so much on the principal content that they simply do not notice the writing itself. If they do not notice the impact of a hedge within the sentence, it is difficult to recreate the desired level of certainty in their own texts.

Using appropriate hedges or boosters is also difficult because, as shown in the taxonomy of hedges and boosters above (Table 7.1), the range of possibilities is quite broad. Modals are a common means of hedging, but the scientist's first language may not have modals, or the same range of modals. The possibility of combining modals or hedging devices further adds subtlety, as in something "*may seem* to indicate" or something "*could possibly* be explained by" something else. Other possibilities of word choice, adverbials, and imprecise numbering (e.g. more than half, a few, a good proportion) can confound the non-native speaker who is attempting to appropriately place the level of claim. The difficulty is recognized in the useful and widely-used guidebook for non-native-English speaking scholars, *Academic Writing for Graduate Students* (Swales and Feak 2004). They devote seven pages to the linguistic elements of hedges and boosters.

Before leaving this discussion of claims and how to linguistically construct them there is some recent geopolitical research on making claims. There is some evidence that there is sensitivity on the part of journal reviewers as to who can make what claims. This sensitivity particularly affects scholars who are situated outside the U.S and the U.K. European scientists face difficulties in "attempting to make a significant contribution in their papers because of reviewers' insistence on geographic limitation of the work" (Lillis and Curry 2010, p. 139). For example, journal reviewers questioned a Spanish psychologist and a Hungarian educator about the relevance of their work to an English-language journal, as it seems that reviewers assume that "studies done outside the Anglophone world… are not considered relevant" to the Anglophone context (Lillis and Curry 2010, p. 142). This creates great difficulty in trying to establish a claim about their work: If they focus on their local context in their work, they are accused of being "parochial" or limited in their scope; if they note the non-Anglophone context of their study, then they find that they are thwarted in making claims of universal relevance or status (Lillis and Curry 2010). In other anecdotal evidence, a Mexican biologist who argued for the first appearance of a disease element in shrimp was told by a reviewer that his data was "untrustworthy" and thus his claim was "exaggerated" (Dr. Jorge de la Rosa, personal communication, 2008). A Korean scholar argues that big ideas cannot come from his colleagues because their claims "may be prejudged [by Western referees] in light of [Koreans'] assumed lower status" (Hwang 2008, p. 106). The negotiation of the intended level of claim and the level that is accepted by the journal reviewers may be complicated by status, geography, language and the scientific content.

7.2 Passives and People

Academic writing is often recognized and even criticized for the depersonalized manner of writing that is expected. The passive voice is included as a typical feature of this kind of writing. In fact, the passive was used three times in the previous two sentences: "writing is recognized", "depersonalized manner is expected", and "passive voice is included". In this kind of grammatical structure, there isn't a person who is doing something; instead, something "is done" by people or forces unnamed. Who is recognizing the writing? Who is criticizing the writing? Who is including the passive voice as typical? The use of passives has grown consistently in science writing over the past 300 years (Banks 2008). And while its use in physics journals used to outnumber its use in biology journals, by the middle of the twentieth century, it was used equally commonly in both the hard sciences and life sciences.

The passive as a type of sentence has several advantages for science writers. It is commonly pointed out that the passive voice "helps the writer to maintain an air of scientific impersonality" (Cooray 1967, cited in Banks 2008, p. 139). In a cynical manner, the impersonality serves as a kind of insurance policy against research difficulty. As one microbiologist explained:

> Everybody wants to put things in the third person. So they just say, 'it was found that'. If it's later shown that it was wrong, don't accept any responsibility. '*It* was found… I didn't say I *believed* it. It was found.' So you sort of get away from yourself that way and make it sound like these things just fall down into your lab notebook and you report them like a historian…' (Gilbert and Mulkay 1984, p. 58).

The syntactic construction of the passive voice sentence foregrounds the activity or event, and not people, which serves Methods sections especially well. In reporting an experimental method, what is important is that, for example, colorant was added. The fact that it was the researcher or a lab assistant who added the colorant is not important. Using the passive removes the researcher from view so that the reader can focus on the activity or idea being presented. The exception to this is when there is an unusual modification to a method that the researcher needs to acknowledge. In this case, sentences often begin with the person pronoun "we", followed by the activity, as in, "We minimized amoeba transformation by completing all zooplankton feeding trials".

Commonly, passive sentences in scientific texts do not have a *by*-phrase, i.e. who or what was performing the action is not stated (Banks 1990). It would be unusual to see (emphasis added): "Colorant was added to the substance **by the research team**" or "Colorant was added to the substance **by automatic infusions**". Applied linguist David Banks analyzed the sentences in 11 oceanography articles that *did* have a *by*-phrase. He found that those clauses are likely to be the abstract or physical cause of the event (e.g. *by coastally trapped waves, by independent estimates*) or the *ing* form of a verb acting as instrument (e.g. *by omitting density gradients*). This latter is seen in the quote above, "we minimized… by completing all zooplankton feeding trials." Notably, "human agents are fairly rare" in science research papers (Banks 1990, p. 341).

Passives are used with particular verbs. They are especially frequent for "material processes", i.e. verbs of activity and change (Martínez 2001). The second most common category of verbs used in the passive is "relational processes", i.e. the verb "to be". It is typical in scientific writing that things happen (material) or things simply are (relational). The researcher is not seen to be active in the research activities. This style of writing allows the writer "to retreat to the background, to foreground findings, to convey the impersonal and factual character that results from the use of incongruent structures of low negotiability" (Martínez 2001, p. 242). According to one applied linguist, together, these elements create the impression that "the facts are speaking for themselves" (Martínez 2001, p. 242).

Despite the common use of passives and the infrequent use of the *I* and *we* in scientific writing, these pronouns are not entirely absent. In most scientific papers across most disciplines, *we* is used in three conspicuous places: Introduction, Discussion, and possibly Method. In the Introduction, writers typically state the purpose or focus of the paper with *we*. For example, "In this paper we present…" or "In this *Letter* we experimentally test…" In the Discussion section, *we* is used especially when referring to important contributions of new knowledge. For example, "We have created polarization-entangled photon pairs" or "We found that a dimmed but visible computer monitor adjacent to the optical table gave negligible background coincidence counts." In the Method section, as mentioned above, *we* is used to distinguish commonly accepted methods from those that are altered by the researchers (Swales and Feak 2004), as in, "To change the relative phase of the local oscillators we tilt a birefringent quartz plate". It is also common to use *we* when describing a series of calculations, such as: "In this case, we can write f(ɥ,f) …. If we drop the two final terms…. We obtain the inequality…."

In summary, writers of scientific articles tend to put themselves in the background in the text. Epistemologically, things and events foregrounded, and the agents, when they are named, are rarely human. Material processes (verbs that indicate action and transformation) are most often put into the passive voice and this sustains the backgrounding of the researcher. Most often, in scientific research papers, the focus is on the scientific activity and not the scientists. However, scientists deliberately bring attention to themselves through the pronoun *we* when they indicate their contributions. After all, scientists typically seek to bring new knowledge to their discipline, and so they highlight their own research purpose in the Introduction and their new contribution in the Discussion. Thus the research paper is a careful construction where the most of the activity revolves around the research process, and only occasionally and strategically are the researchers made visible.

7.3 Noun Structures and Nominalizations

A notable element of scientific writing is the presence of dense noun phrases that condense a great deal of information into a small space. In the everyday world, people do things that involve other participants, whether those participants are animate or inanimate. For example, in microbiology one might say, "We transferred DNA from agarose gels to Hy-bond-N nylon membranes," (Corbett 1992, p. 43). The researchers "we" performed the action of "transferring" the DNA from one location to another. However, this way of reporting the researchers' actions is not common in science papers. Placing the researchers in such a prominent and visible position is not typical, as we discussed in Sect. 7.2 regarding passive sentences.

The focus in science writing is generally not on what the researchers do but on the processes themselves. Thus a more common way of writing about transferring the DNA would be, "Transfer of DNA from agarose gels to Hy-bond-N nylon membranes *was done* in accordance with suppliers instructions," (emphasis added, Corbett 1992, p. 43). This is an example of creating a noun ("transfer of DNA") from an action ("we transferred the DNA"). This transformation of an action into a noun is called nominalization and the resulting noun phrases are termed grammatical metaphors. They serve to "compactly convey… technical messages to small groups of highly trained readers in a specialized research field" (Gross et al. 2002, p. 167). It is a compressed form of writing that now appears in "virtually every sentence in present-day written academic text" (Biber and Gray 2010, p. 17).

This procedure of nominalization can be seen in the writing of Isaac Newton in 1704. Linguists Halliday and Martin (1993) examine a single paragraph written by Newton concerning the optics of the eye. The original says:

> If the Humours of the Eye by old Age decay, so as by shrinking to make the *Cornea* and Coat of the *Crystalline Humour* grow flatter than before, the Light will not be refracted enough, and for want of sufficient Refraction, will not converge to the bottom of the Eye but to some place beyond it, and by consequence paint in the bottom of the Eye a confused Picture, and according to the Indistinctness of the Picture the Object will appear confused. ….For those Convex glasses supply the defect of plumpness in the Eye…. (Newton 1704, cited in Halliday and Martin 1993, p. 7).

Halliday and Martin (1993, p. 7) point out how words that are verbs become transformed into nouns in Newton's text:

> will not be refracted enough….for want of sufficient Refraction
> paint a confused Picture….according to the Indistinctness of this Picture
> make the Cornea grow flatter….supply the defect of Plumpness.

When an action becomes frozen as a noun, then focus on manipulation or activity is removed. In Newton's passage above, the action of refracting light by some means becomes frozen as the state of refraction. Material processes that would otherwise indicate activity become nouns in this linguistic structure. Further, words that had separate functions become collapsed as modifiers of the noun.

Nominalization structures allow "the expression of complex, information-rich utterances in a compact form" (Hanauer and Englander 2013, p. 22). The scientist may create new nominalizations and definitions as he or she writes a paper in much the same way Newton did 400 years ago. For example,

> In the process of spontaneous parametric *downconversion*, photons from a pump field are converted…. The reverse process, or *upconversion* of these pairs back to single photons

in a second crystal, is also possible. Here, we present experimental measurements of the *upconversion rate* with a controlled time delay…. (O'Donnell and U'Ren 2009, p. 603).

In this example, the term 'downconversion' which is a known term is contrasted with a new term, 'upconversion'. Having established what 'upconversion' is, it can then be used as a modifier for 'rate' so that the writers next speak of the 'upconversion rate'.

Within a discipline, most of the nominalized terms are already known and "the vast majority of the terms are already established within the discourse community at the time of writing" (Banks 2008, p. 135). Thus when experts read about "downconversion" and its opposite "upconversion", "a pump field" or "controlled time delay", they recognize it as an easily understood single conceptual entity. It serves as a kind of abbreviation that has been called telegraphic (Hanauer and Englander 2013).

Both the novice reader and non-native-English-speaker may find these structures very difficult. The novice reader "would need to literally decipher the various elements of the noun phrase and relate them to real world" (Hanauer and Englander 2013, p. 22). To produce these structures, the writer must take active elements and condense them down into these complex noun phrases. For example, a published medical paper with little condensing stated, "Many patients were followed for up to 2 years and no new PKDL cases were detected after the 6-month period" (ElMalik and Nesi 2008, p. 94). That was written by a non-native-English speaker, and it might have been written as, "Follow-up studies for up to 2 years detected no new PKDL cases after the 6-month period". The rewritten versions use only one verb, "detected", and all other elements are nominalized, greatly condensing the text (18 vs. 21 words; 2 clauses vs. one clause). Such a compact and condensed writing style can be difficult to master.

Nominalizations have been used in both the physical and biological sciences, beginning as early as Newton in the early eighteenth century (Halliday and Martin 1993). However, their use has increased steadily in the twentieth century (Banks 2008), making this structure a key feature of scientific writing.

7.4 Summary

Creating knowledge in a scientific paper is a complex linguistic and social process. The scientific paper is not a linear story of narrative scientific discovery because it rarely reports the discovery process based on a true-to-life recounting of events with all the research's twists and turns. Instead, it is a document that the reports the discovery process in a dense, compact manner that has evolved to meet the needs of the scientific community. The paper serves to persuade the expert reader of the new knowledge with the greatest possible degree of generalization.

Scientific papers foreground the activity of science creation and largely background the role of the researchers. This presentation is accomplished through specific linguistic resources such as hedging and boosting in making claims, passives and pronouns, and complex noun phrases. The genre of the scientific research paper is instantly recognizable and becomes an easily comprehensible text for the expert reader in a shared science community. In the next chapter, the linguistic and epistemological structures of the prototypical research article's Introduction, Method, Results and Discussion are discussed.

References

Banks, D. (1990). Agents and instruments in scientific writing. *IRAL: International Journal of Applied Linguistics in Language Teaching, 28*, 336–347.

Banks, D. (2008). *The development of scientific writing: Linguistic features and historical context*. London: Equinox.

Bazerman, C. (1988). *Shaping written knowledge*. Madison, WI: University of Wisconsin Press.

Biber, D., & Gray, B. (2010). Challenging stereotypes about academic writing: Complexity, elaboration, explicitness. *English for Academic Purposes, 9*(1), 2–20.

Cho, D. W. (2009). Science journal paper writing in an EFL context. *English for Specific Purposes, 28*, 230–239.

Corbett, J. (1992). Encoding nature: Writing science in English. *Microbiology Letters, 100*, 39–44.

Crismore, A., & Vande Kopple, W. J. (1997). Hedges and readers: Effects on attitudes and learning. In R. Markkanen & H. Schroder (Eds.), *Hedging and discourse: Appraoches to the analysis of a pragmatic phenomenon in academic texts* (pp. 83–114). Berlin: Walter de Gruyter.

ElMalik, A. T., & Nesi, H. (2008). Publishing research in a second language: The case of Sudanese contributors to international medical journals. *English for Academic Purposes, 7*, 87–96.

Englander, K. (2009). Transformation of the identities of nonnative English speaking scientists. *Journal of Language, Identity and Education, 8*, 35–53.

Flowerdew, J. (1999). Problems in writing for scholarly publication in English: the case of Hong Kong. *Journal of Second Language Writing, 8*(3), 243–264.

Gilbert, G. N., & Mulkay, M. (1984). *Opening Pandora's box: A sociological analysis of scientists' discourse*. Cambridge: Cambridge University Press.

Grabe, W., & Kaplan, R. B. (1997). On the writing of science and the science of writing: Hedging in science text and elsewhere. In R. Markkanen & H. Schroder (Eds.), *Hedging and discourse* (pp. 151–167). Berlin: Walter de Gruyter.

Gross, A. G., Harmon, J. E., & Reidy, M. S. (2002). *Communicating science: The scientific article from the 17th century to the present*. West Lafayette, IN: Parlor Press.

Halliday, M. A., & Martin, J. R. (1993). *Writing science: Literacy and discursive power*. Pittsburgh: University of Pittsburgh Press.

Hanauer, D. I., & Englander, K. (2013). *Writing science in a second language*. West Lafayette, IN: Parlor Press.

Hwang, K. (2008). International collaboration in multilayered center-periphery in the globalization of science and technology. *Science, Technology and Human Values, 23*, 101–133.

Hyland, K. (2000a). Hedges, boosters, and lexical invisibility: Noticing modifiers in academic texts. *Language Awareness, 9*(4), 179–197.

Hyland, K. (2000b). *Disciplinary discourses: Social interactions in academic writing*. Harlow, England: Longman.

Hyland, K. (2005). Stance and engagement: A model of interaction in academic discourse. *Discourse Studies, 7*(2), 173–192.

Knorr-Cetina, K. (1981). *The manufacture of knowledge: An essay on the constructivist and contextual nature of science*. Oxford: Pergamon.

Kreutz, H., & Harres, A. (1997). Some observations on the distribution and function of hedging in German and English academic writing. In A. Duszak (Ed.), *Culture and styles in academic discourse* (pp. 181–202). Berlin: Mouton de Gruyter.

Latour, B., & Woolgar, S. (1986). *Laboratory life: The social construction of scientific facts.* Princeton, NJ: Princeton University Press.

Lillis, T., & Curry, M. J. (2010). *Academic writing in a global context: The politics and practices of publishing in English.* London: Routledge.

Martínez, I. A. (2001). Impersonality in the research article as revealed by analysis of the transitivity structure. *English for Specific Purposes, 20*, 227–247.

Myers, G. (1985). Texts as knowledge claims: The social construction of two biology articles. *Social Studies of Science, 15*, 593–630.

O'Donnell, K., & U'Ren, A. (2009). Time-resolved up-conversion of entangled photon pairs. *Physical Review Letters, 102*(12), 602–606.

Pinch, T. (1985). Towards an analysis of scientific observation: The externality and evidential significance of observational reports in physics. *Social Studies of Science, 15*, 3–36.

Swales, J. M., & Feak, C. (2004). *Academic writing for graduate students: Essential tasks and skills* (2nd ed.). Ann Arbor, MI: University of Michigan Press.

Vassileva, I. (1997). Hedging in English and Bulgarian academic writing. In A. Duszak (Ed.), *Culture and styles in academic discourse* (pp. 141–157). Berlin: Mouton de Gruyter.

Zerbe, M. J. (2007). *Composition and the rhetoric of science: Engaging dominant discourse.* Carbondale, IL: Southern Illinois University Press.

Chapter 8
Writing the Five Principal Sections: Abstract, Introduction, Methods, Results and Discussion

Abstract The five-part structure of research papers (Introduction, Method, Results and Discussion plus Abstract) serves as the conceptual basis for the content. The structure can be considered an "hourglass" or a "conversation" with predictable elements. Each section of the paper has its own common internal structure and linguistic features. The structures and features are explained with examples from published papers in a range of scientific disciplines. Exceptions to the four-part structure are also discussed.

Keywords IMRD • General to specific • Content • Moves • Claims • Introduction • Methods • Results • Discussion • Rhetoric • How to write

Typically, scientific research papers have five principal sections: Introduction, Method, Results and Discussion, plus an Abstract. Together this is referred to as the IMRD (pronounced "im-rad") format for its initials. The IMRD format is especially common for reporting quantitative, experiment-based research. Qualitative and non-experimental research papers are somewhat different in the middle sections of the organizational structure. Analysis of both types of papers is presented in this chapter. Papers which are reviews, state-of-the-art surveys, rebuttals or opinions will not typically be structured in this way and they are not directly addressed in this chapter. However, the evidence that is presented in this chapter is based on the analysis performed by applied linguists who have amassed and analyzed large numbers of scientific papers that report research in order to characterize the way that published scientific research papers are written.

8.1 Structures for the Research Paper

Two ways of thinking about the structure of the research paper are as an hourglass and as part of a conversation. By linking together these two concepts, both the purpose and the linguistic structure of each section of research papers can

SpringerBriefs in Education, DOI: 10.1007/978-94-007-7714-9_8, © The Author(s) 2014

be examined. An overview of the two concepts is presented here, and a detailed description of each section follows below.

8.1.1 The IMRD Hourglass

An hourglass is an old-fashioned piece of technology that was used to count one hour (Fig. 8.1a). Its shape bears a striking resemblance to the structure of the four sections of the IMRD research paper (Fig. 8.1b), so we can say that the research paper is shaped like an hourglass.

The Introduction section typically begins with wide, general statements about the importance of the area of study, and then progressively reports what is known about the area, what is still not known, and then states the specific focus of the paper. The Introduction moves from content that is general, to content that is specific. Visually, this is pictured as moving from wide to narrow, as in the top portion of the hourglass. The middle sections of the research paper continue the narrow focus of the specific research being reported in the Methods and Results. This is pictured as the narrow central area of the hourglass. The Discussion usually begins with comments about the specific research and progressively connects the findings to the wider literature in the field. This move from specific content of the research investigation to the larger field is pictured as a moving from narrower to wider, as seen in the bottom portion of the hourglass.

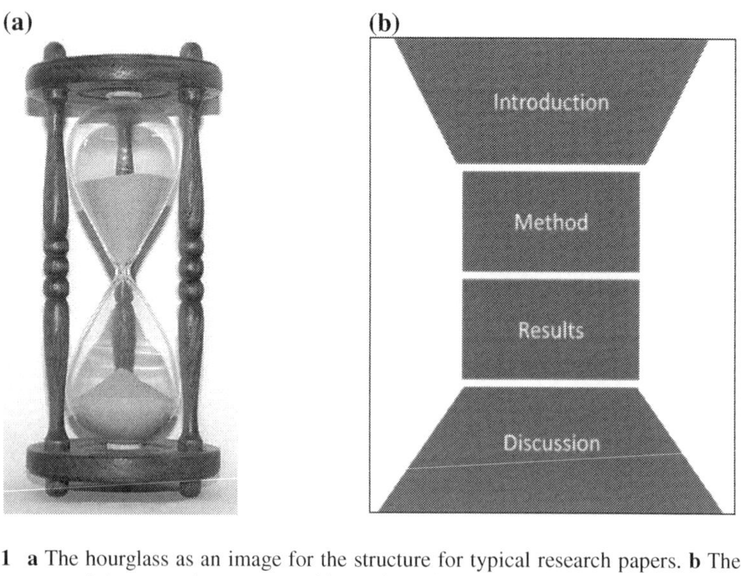

Fig. 8.1 **a** The hourglass as an image for the structure for typical research papers. **b** The shape of the contents of the research paper resembles an hourglass

The image of the hourglass captures the movement of ideas within a research paper. The ideas begin by referring to the larger scientific area under study, narrow the focus to the specific study being reported, and then broaden the comments to include the larger scientific community again. The Discussion at the bottom serves as an inverse commentary on the Introduction. The overall flow of the research paper can be visualized in this way.

8.1.2 The Research Paper as Conversation

Another way of thinking about a research paper is that it participates in a conversation. The conversation works in the following manner. Contributions to a topic of inquiry have already been published. A research paper will add to the knowledge, and thus add to the conversation about that area of inquiry. Other researchers will continue the topic of conversation, in part by referring to your contribution. Thus the research paper responds to what has been written, and as others continue the conversation, your contribution will be referenced. This is a conversation on the topic that has preceded this paper, and a conversation that will be continued by others. The paper that any scientist writes is part of the conversation, and if that paper 'speaks well', others will refer to it in subsequent turns of the conversation.

To speak well, each section of the research paper has a purpose that demonstrates to future readers that the contribution should be taken seriously. As each section of the research paper is discussed below, the manner in which each section supports a paper's ability to speak well is also discussed.

8.2 The Introduction

Not all scientists read the Introduction of research papers! Some scientists say that Introductions don't actually include any new information, so there is little point to wasting time reading them, since their time is always limited. However, Introductions play a crucial role in establishing the work, and it is a section that is read by journal reviewers when they are evaluating the suitability of a paper for publication. Ethnographic studies of scientists writing research papers indicate that it requires the most composing time and rhetorical effort because it serves a strategic importance.

Thinking of the research paper as part of a conversation, the Introduction makes three important points:

(1) There is an area of study that we—peers who read this journal—agree is important;
(2) There are certain things that are known about that area of study and I can show you that I know about that prior work;
(3) There is still something about this that we don't know.

The third item—there is something that we don't know—is sometimes referred to as a gap in the field. The existence of the gap is presented as the motivation for the present study. In the logic of the conversation, the scientist fills the gap by performing the particular research study that is presented in this research paper. There is typically an overt statement of purpose of the paper, and in the statement of purpose, it is clear that the scientist is about to make a new contribution to the field's knowledge through the work presented in the paper. For example, consider the Introduction shown in Fig. 8.2.

The authors open by stating the scope of the topic: huge numbers of photographs are published in the press and posted by ordinary people every day. The next sentence states the area that is "important": to understand how people perceive pictures. The topic and the area of importance are established in the first paragraph. The second paragraph specifies further how photographs are the consequence of effects of the camera, includes figures to make the point, and refers to the work of others. This paragraph presents existing knowledge about the topic. The third paragraph presents the problem in the existing knowledge, which is that the perception distortion of some effects occurs even though they "are not caused by physical distortion". Since the effects are not caused by physical distortion, then researchers "must consider perceptual mechanisms and people's viewing habits." The authors then point out that such considerations are the purpose of their paper.

In this Introduction, the authors participate in a conversation in their field regarding vision perception. They say the area is important, and it has been studied to some extent but not everything that is necessary to know about how people perceive pictures has yet been determined. Thus the purpose of their paper is to contribute to something new to the knowledge about the area. This should allow subsequent researchers to build on this knowledge.

Introduction

Every day the Associated Press publishes ~3,000 new photographs, and Facebook users post nearly 250 million (Associated Press, 2009; Shaffer, 2011). Clearly, people rely on pictures to communicate with one another, so it is important to understand how people perceive pictures in typical viewing situations.

Photographers, cinematographers, and computer-graphics engineers create pictorial effects in various ways. For example, photographs of scenes captured with short-focal-length lenses appear expanded in depth, while those captured with long lenses appear compressed. These effects can be seen in still photographs and video. In the latter, the technique creating the effect is called a "dolly-zoom" shot. Figure 1a shows two example photographs from the website Flickr. On the left, the goat looks stretched in depth; on the right, the pitcher and batter appear to be much closer to one another than they actually are. Figure 1b shows two example frames from a dolly-zoom shot in the movie *Goodfellas*. Objects through the window appear much farther away in the frame on the left (from early in the scene) than in the frame on the right (from later). Figure 1c shows how depth compression and expansion can also affect the appearance of a face. Long lenses can make a person look smarter, more attractive, and less approachable; short lenses have the opposite effect (Perona, 2007).

The apparent expansions and compressions in depth are often called perspective distortion, as if these effects are due to a distortion in the physical projection from the scene to the film plane. The effects occur, however, when the projections are geometrically correct. For example, Figures 2 and 3 contain computer-generated (CG) videos of dolly-zoom shots of a kitchen and a face. Both were created using accurate 3D models and correct perspective projection. Thus, the perceptual effects are not caused by physical distortion in the projections. To explain them, one must consider perceptual mechanisms and people's viewing habits, and that is the purpose of our paper.

Fig. 8.2 Introduction section of paper in field of vision research (Cooper et al. 2012, p. 1, reprinted with permission)

Another way that Introductions have been analyzed is through a move structure. There are three typical "moves" or logical and rhetorical concepts in an Introduction. The moves of an Introduction were first detailed by John Swales (1990) who has famously used an ecological metaphor for "creating a research space".

- Move 1 establishes the research territory.
- Move 2 establishes a niche. A niche in ecology is "a particular microenvironment where a particular organism can thrive; in our case, a niche is a context where a particular piece of research makes particularly good sense" (Swales and Feak 2004, p. 244).
- Move 3 occupies the niche.

The moves, like the shape of the top section of the hourglass, go from the general to the specific in their focus. Figure 8.3 expands on the content of each of the moves. It details how the move is accomplished, in other words, how the writer establishes the research territory. This is done by showing that that general research area is "important, central, interesting, problematic, or relevant in some way" (Swales and Feak 2004, p. 244). Note that this content is sometimes optional, as not all journals or disciplines require that the research territory be established in this way. However, all papers establish the territory by "introducing and reviewing items of previous research in the area" (Swales and Feak 2004, p. 244). This is an obligatory element of Move 1. The elements of Move 2 and Move 3 are also detailed in Fig. 8.3. See Boxes 8.1 to 8.4 for examples of how scientists create the moves through their words in their papers.

Examples of how scientists create Move 1a in their papers are:
A precise chronological framework…is critical for ….
A fundamental question in the study of … is the problem of ….
The characterization of … is of great importance since ….

Box 8.1 Examples of how scientists establish the importance of a study area in the Introduction

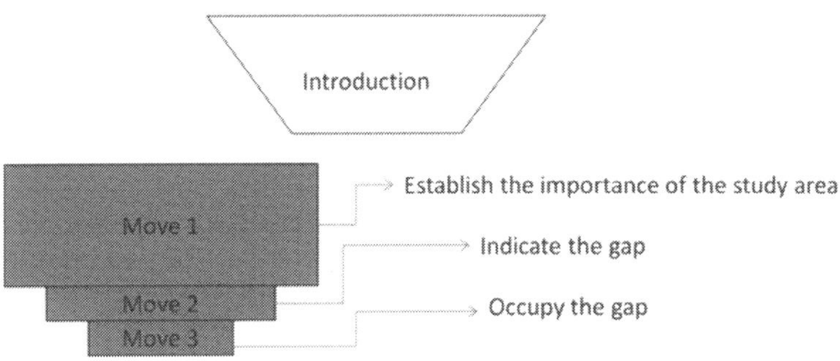

Fig. 8.3 Elements of Move 1 as part of the Introduction

Examples of how scientists create Move 1b in their papers are:
It has previously been proposed that....
Based on these findings [of other researchers], it has been argued that....

Box 8.2 Examples of how scientists introduce and review previous research in the area. Often the present perfect tense is used: "has been proposed", "has been argued".

Examples of how scientists create Move 2 in their papers are:
Unfortunately, many ...lack .....
While it has been shown that...the basis for...remains largely unknown.
However, ...
It is of great interest to study whether this ... has ... or ....
Little is known about how these factors influence....

Box 8.3 Examples of how scientists indicate a gap or extend the knowledge of the existing research

Examples of how scientists create Move 3a in their papers are:
In this study we employed several analytical techniques....We obtained....In addition, we report....
Alternatively, we propose that...Here we present examples of such features.....
A key step in understanding...is to understand....This is the goal of the present paper. We focus here on....
We therefore propose an approach based on....
Based on the above, the aim of this work was to.....
The present study was designed to investigate the effect of ... on ....

Box 8.4 Examples of how scientists occupy the gap by outlining the purpose(s) or stating the nature of the present research

It is interesting to note the way that Move 3 statements are written; that is, the sentences that state the purpose or nature of the research. The word "we" is often used, as in "we focus here on" or "we present". Some popular instructions on writing science claim that the writer should never be obvious in the text and that use of "I" or "we" is discouraged. Some fields follow this style, and it can be seen in Box 8.4 in the phrases "the aim of this work was" and "the present study was designed". These statements put "the aim" and "the present study" in the subject position and avoid using "we". However, as noted in Chap. 7, Sect. 7.2, in many science fields, it is entirely appropriate for authors to emphasize exactly what their contribution is. Using "we" to make that contribution clear is very typical (Box 8.4).

To write an effective Introduction, emulating well-respected research papers in the field can be helpful. To aid in understanding how those papers are structured, the image of the hourglass and idea of being part of a conversation serve as a template. To summarize, then, the Introduction is typically written through the three moves, where the content narrows from establishing the importance of the area of study and background about the field, to specifying the purpose and focus of the present research. Visually, this movement is captured in the top portion of the hourglass as it narrows. When considering the research paper as part of a conversation, the content of the Introduction establishes the scientist's credibility to speak about the research. This section of the paper indicates that the scientist knows what the important issues in the field are, what has been said about them, and can identify a significant issue that still needs to be addressed. The final element of the Introduction states how the scientist will address that issue. It may also include how the rest of the paper (this contribution to the conversation) is organized.

Before leaving this discussion of research paper Introductions, it is important to note that not all fields and not all journals within a field strictly follow this format. Some will have a separate background literature section in addition to a brief review in the Introduction. Some may not have a clear statement of the gap in the existing research, although rhetorically it does seem desirable. Other variations are inevitable. Some may have additional sections that describe the geographic site of a study or provide taxonomic information about the subject of study. As discussed in Chap. 5, the research paper is not a static form. It is a genre that has evolved to meet the needs of scientists at specific times in specific fields, so changes and variations will continue to occur. Nonetheless, the foregoing discussion of Introduction sections of research papers represents the sophisticated analysis from linguists and text analysts of what can be considered the common characteristics across fields and across journals that publish research papers in English.

8.3 Materials and Methods Sections

The classic presentation of a research paper's Methods section or Materials and Methods section is seen most commonly in experimental research papers. As will be discussed below, non-experimental papers do not have a section quite like this. Non-experimental fields include many earth and space sciences such as geophysics, oceanology, and astronomy. These latter fields seek to explain existing phenomena, using models typically. The middle portion of these kinds of research papers presents the theoretical basis used for obtaining the results. In contrast, experimental research reports on action taken by the scientist, and the resulting activity is sought to be explained. The Methods section provides a reporting of what and how the experiment was conducted.

Regardless of whether the paper is in the experimental or qualitative field, the concepts of the hourglass and the conversation apply. Within the hourglass image (Fig. 8.1a and b), this section is the upper one of the two narrow, middle sections. It extends directly below the Introduction's Move 3 where the purpose or nature of the present research is stated. This middle section is the same "width" as the narrowest point of the Introduction because it extrapolates directly from the specific statement of purpose of the research. Visually, this section simply deepens the immediately previously stated purpose by elaborating how the research was conducted.

Considering the research paper as part of the conversation of the field, the Methods section establishes that the scientist has performed the research in a credible manner. This section demonstrates that the scientist knows the procedure or procedures that are acceptable for obtaining evidence in the field. The scientist demonstrates that those procedures were performed properly. Where the scientist faced decisions regarding available procedures, those decisions were made for appropriate reasons. Where modifications were made to accepted procedures, they are noted and were done properly. In summary, the Methods section assures the reader that the findings that are reported later can be trusted because the data on which the findings are based were produced an acceptable manner.

Methods sections in experimental papers, or the theoretical assumptions of the work that underlies non-experimental papers, are intended to allow for replication of the work. The amount of detail varies with the field and the specific journal. In some biological fields, the procedures are so standardized that processes are referred to by an author's name and date. For example in a study concerning optimal temperature conditions for halibut fish, the authors wrote:

> The upper incipient lethal temperature (UILT) was studied with the method developed by Fry (1964) and with the criterion of Kilgour et al. (1985)…. The value of the high incipient lethal temperature of the juveniles was analyzed with USEPA (1986) (Spearman-Karber method). (Esquer Mendez et al. 2010, p. 1016)

In the description above of the treatment of fish, it is assumed that the reader will know the method of Fry, the criterion of Kilgour et al. and the Spearman-Karber method. Further elaboration is not required.

Other fields require a much longer explanation of what activities were performed, with what equipment and in what manner. Sometimes, a rationale for the method is required. For example, in a study of sea organisms, the authors needed to demonstrate that they lessened a common problem of the organisms transforming themselves before they can be studied. Their Methods stated (emphasis added):

> **Since some toxic flagellated vegetative cells** in a given population **are induced to transform** to amoeba shortly after disturbance, … **we minimized amoeboid transformation by** completing all zooplankton feeding trials…. In all treatments containing *P. piscicida*, the dinoflagellates were added **by slowing pouring** to containers…. (cited in Penrose and Katz 2001, p. 330)

The authors identify a problem situation of the target cells transforming themselves. Note that the present tense "are induced" is used to describe the problem, indicating that this is a common phenomenon and not limited only to their study.

As noted earlier, authors typically indicate something special, important, or non-typical in Methods by using "we". These authors show that "we" did something to minimize the problem "by" doing particular things. They switch to the past tense to indicate that what they did was particular to their research method. Common procedures are generally reported in the past tense, with no personalization of the authors performing the action, and the passive voice is common. The authors indicated a practice was common by saying "the dinoflagellates were added…." This short excerpt shows many common linguistic characteristics of a Method section.

Papers where no experiments are performed need to articulate the assumptions of the study. The assumptions may include how a physical location was defined, what kind of measurements were collected and how the data were analyzed. To create models, the mathematical principles and formulations for calculations are presented (Fig. 8.4).

Changing technology is having an impact on how journals expect the Method sections to be written. In some cases, Methods in the printed text are reduced to little more than a paragraph, and an extended explanation is provided as a link on the journal's website. Some journals ask for a video of the method that is posted to accompany the published paper. Some journals simply change the positioning of Methods by requiring the section, but it appears after Results and Conclusion. All these innovations are part of the evolving genre of the scientific paper. The changes are the result of the demands of the journal publishing houses, new opportunities for new media, and, perhaps, changing priorities of the scientist-readers as well.

The integral equation for horizontal magnetic dipoles in y-direction following Perez-Flores et al. (2001), we get:

$$\sigma_{a,y}(r_1,r_2) \cong -\frac{16\pi s}{m_y \mu_0}\int_v G^0_{H_y}(r,r_2)\cdot E'_{h_y}(r,r_1)\sigma(r)d^3r \qquad (8)$$

where;

$$E'_{h_y}(r,r_1) = -\frac{\mu_0 m_y}{4\pi}\left\{\left[\frac{1}{\rho_1^2} - \frac{z+h_1}{\rho_1^2|r-r_1|} - \frac{2(y-y_1)^2}{\rho_1^2} + \frac{2(z+h_1)(y-y_1)^2}{\rho_1^4|r-r_1|} + \frac{z+h_1}{\rho_1^2|r-r_1|^3}\right]i + \left[\frac{(x-x_1)(y-y_1)}{\rho_1^2}\left(\frac{2}{\rho_1^2} - \frac{2(z+h_1)}{\rho_1^2|r-r_1|} - \frac{z+h_1}{|r-r_1|^3}\right)\right]j\right\}, \qquad (9)$$

$$G^0_{H_y}(r,r_2) = \frac{1}{4\pi}\left\{\left[\frac{1}{\rho_2^2} - \frac{z+h_2}{\rho_2^2|r-r_2|} - \frac{2(y-y_2)^2}{\rho_2^2} + \frac{2(z+h_2)(y-y_2)^2}{\rho_2^4|r-r_2|} + \frac{z+h_2}{\rho_2^2|r-r_2|^3}\right]i + \left[\frac{(x-x_2)(y-y_2)}{\rho_2^2}\left(\frac{2}{\rho_2^2} - \frac{2(z+h_2)}{\rho_2^2|r-r_2|} - \frac{z+h_2}{|r-r_2|^3}\right)\right]j\right\}. \qquad (10)$$

The product of equations (9) and (10) does not give a simpler equation as in equation (7), but it is not difficult to compute them separately and put them together to obtain $\sigma_{a,y}(r_1,r_2)$.

Fig. 8.4 In a geophysics paper concerning a method for analyzing satellite data to interpret land formations, the method involved stating the assumptions of the mathematical calculations. (Early draft of manuscript, Pérez-Flores et al. 2011)

8.4 Results Section

When we consider the research paper as part of a conversation, the Results tell the reader that something interesting was found as a result of the careful Methods that were implemented. The author needs to highlight the results that are interesting and help the reader to recognize what is important. This section tells the essential story of the paper. A useful way to think about Results comes from answering four questions, (excerpted from Cargill and O'Connor 2009, p. 21):

1. What do my results say? (in two sentences)
2. What do these results mean in their context? (i.e. what conclusions can be drawn from these results?)
3. Who needs to know about these results? (i.e. who specifically forms the audience for this paper?)
4. Why do they need to know? (i.e. what contribution will the results make to the ongoing work in the field? Or, what will other researchers be missing if they don't read this paper?)

The answers to these questions can help to identify what is important to highlight and why. Making these decisions about the data is necessary because Results do not necessarily report *everything* that occurred, and it shapes the significance of the data (Fig. 8.5).

A common assumption is that "the results speak for themselves." In fact, through tables, figures and text, the author directs the reader's attention to what is important and how a result should be interpreted. See Fig. 8.6, and note the two possible captions.

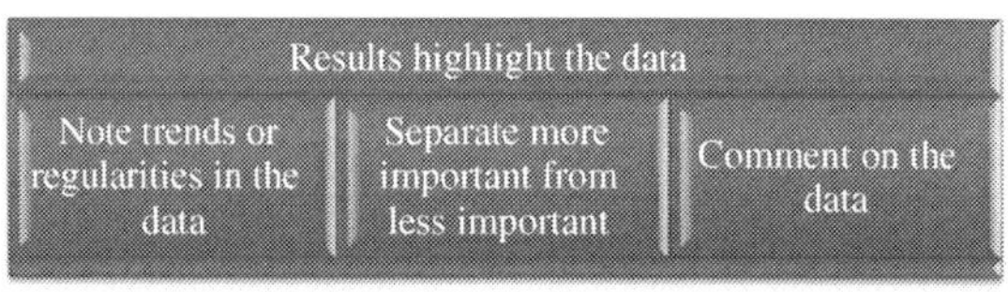

Fig. 8.5 Three activities an author must perform in the Results section in order to highlight the data

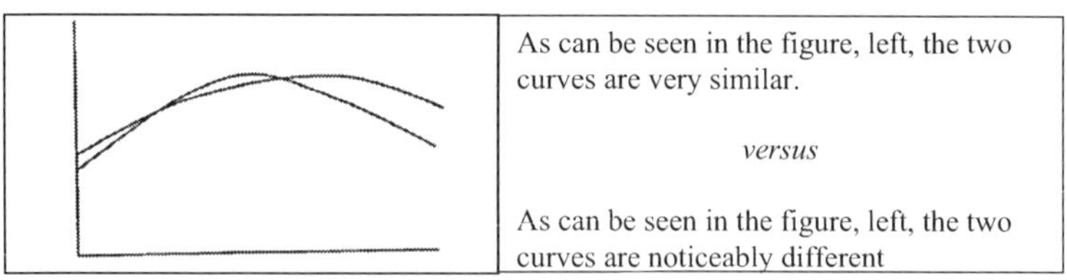

Fig. 8.6 A figure and its caption can direct the reader to the results that are noteworthy (Glassman-Deal 2010, p. 103, reprinted with permission)

The author may identify the curves as being largely similar or largely different. In doing so, the author indicates the preferred interpretation. The later explanation of what produced such "similar" or "divergent" results will be taken up in the Discussion.

This dichotomy of interpretation may seem like an extreme example, but providing interpretative information occurs frequently in Results sections. Note the highlighted words in the text in Box 8.5. The word "significantly" is not necessarily statistically determined except in the last sentence, but the word is used as an adjective to indicate importance. The first finding is compared with a result that is "second best". Later, the authors say the results "seem to support" the fact that "practically all" the nutritional requirements for the sea creature that is being fed different diets.

> The mean total yield of the cultures fed C. muelleri was **significantly** higher than those obtained at the end of the first experiment with the other diets. I. galbana gave the **second best** result followed by the N. oculata and frozen Tetraselmis-fed cultures, which were **not significantly different** from each other but **gave better results** than those obtained with T. suecica….
>
> The results of this second experiment showed that C. muelleri **seems** to support **practically all** the nutritional requirements of P. euryhalinus, **since** none of the mixtures could improve the nutritional value of monospecific C. muelleri cultures. The mean yields obtained in this experiment ranged from 462.92 ± 152.5 organism $L - 1$ (C. muelleri) to 296 ± 114.74 $L - 1$ with the mixture C. muelleri, I. galbana and frozen Tetraselmis sp., and there **were no significant differences** ($P = 0.116$) between diets.

Box 8.5 Scientists use words (highlighted) to indicate their interpretation of the data by emphasizing what is important to what degree and giving reasons for their interpretation of those data.

Scientists use words to emphasize highlights in the data frequently. Note the variety expressions and the emphasis the authors use in extracts from the Results section in other published papers in Box 8.6.

> Of **great** interest are….
> There was **clearly** no effect of….
> This is a **useful** property that is not included in [other] schemes.
> …indicating that features other than shape **certainly** play a role.
> From Fig. 8.1, **it is obvious** that there is **substantial** individual variability.

Box 8.6 Scientists use expressions in their Results to help the reader understand the significance of the data.

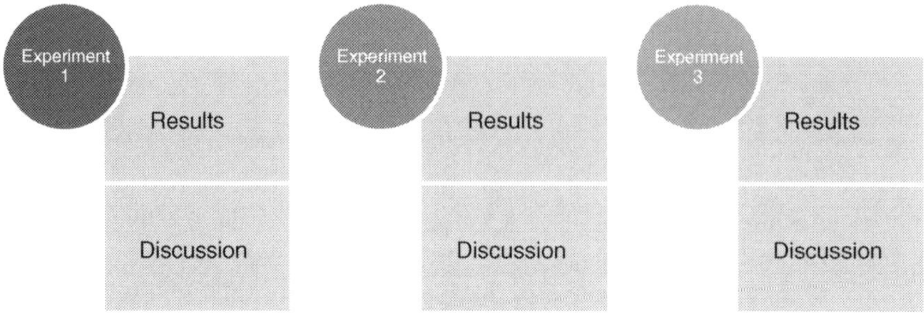

Fig. 8.7 Structure of papers where Results and Discussion are presented together in relation to a series of experiments

Authors point to the findings that they interpret as being important in the study by carefully choosing words which emphasize or downplay the data. At the same time, this careful wording of the Results helps the reader to recognize the significance of the findings, which is further discussed in the Discussion.

Figures and tables are typically central elements of the Results. They should be able to stand alone; i.e. the image and the caption should be clear to the reader without having to read the Results section to understand it. Of course, every figure and table must also be referred to in the text.

The decision whether to present the data as a figure or a table often depends on the information the author wants the reader to notice. If the focus is on trends, a comparison of a few elements, or the "shape" of the data, then a figure is most appropriate. On the other hand, if the purpose is to focus on individual data values, recording raw data, explaining calculations or allowing comparisons of data in many directions, then a table is best. It is not unusual for professional scientists to report that they "read the figures first" in a research paper (James Elder, personal communication, July 20, 2012). If the story that is told through the figures is clear and interesting, then they read the text of the paper.

In this chapter, the Results section has been discussed as one unit within a research paper. However, some scientific fields and journals do not present all the Results together and then, separately, present the Discussion. Some papers separate one set of Results and immediately follow this with Discussion. Then the next Result is presented followed by a Discussion of that data, and so on, through all the relevant results (Fig. 8.7). This is especially common in experimental fields where several inter-related experiments are reported. Papers with this structure typically end with a Conclusion.

8.5 Discussion Section

The Discussion section of a paper presents the new contribution to the field. When thinking of contributing to the conversation, this section explains what the findings mean and how they alter what was known before. When considering the shape of

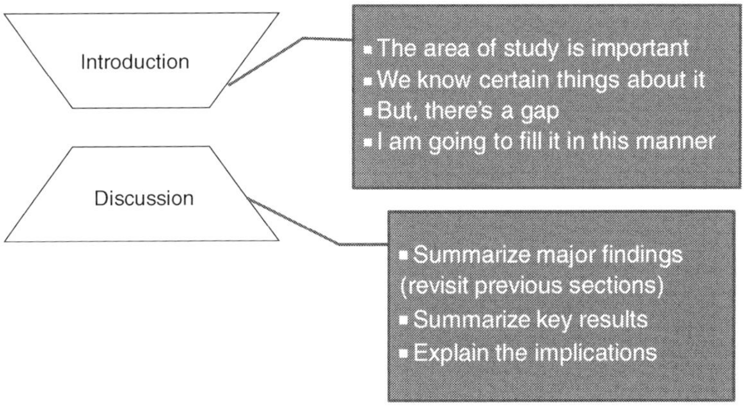

Fig. 8.8 The Discussion section mirrors the Introduction by moving the ideas from the specific findings to the implications for the field; a reflection of the Introduction that begins with the larger field and narrows to focus on the specific study

the research paper, the Discussion serves almost as a mirror of the Introduction (Fig. 8.8).

Recall that the Introduction begins with general statements about the importance of the field and the specific area of study. It recounts what is known about the phenomenon based on the published literature. It presents a gap in the existing knowledge and then moves to a specific statement that indicates the purpose or focus of this particular research paper which fills that gap.

In this progression of ideas, the statements move from the broader more general content, to narrower and more specific content. The inverse is true, typically, of the Discussion section. The specific results of the research are summarized specifically. The results are "discussed" in terms of similarity, contrast and extension with findings and explanations in the published literature. In other words, the specific findings are linked to the larger literature, and so broadening the scope of the findings. Then, the implications of this new knowledge for the field are presented. In some cases, scientists stake a claim to the further research. The content in this section conceptually broadens out. Thus while the Introduction moves from broad ideas to narrow focus, the Discussion moves from the narrow focus of the specific findings of the study to the broader field. In this sense, the Discussion reflects or mirrors the Introduction.

Discussions typically open with a brief recap of the study itself but the central element of this section is making claims about the findings. How important are the findings? How definitive are they? How certain is the scientist of their interpretation or explanation of the findings? These questions are answered through the level of claim that the author uses concerning the work.

Claims can be identified as five different levels or types (Latour and Woolgar 1986) that set on a continuum from speculative assertions to fact-like entities, as discussed in Chap. 7 Sect. 7.1. Type 1 claims are speculative statements and consist of conjectures. Type 2 claims are tentative and offer thoughts on what might reasonably be concluded from the data. Type 3 statements typically use terms of probability and modality (see Fig. 8.9) that serve to question the certainty of the claim. Type 4 claims are explicitly stated as having a definite foundation. Type 5 claims are statements that no longer need discussion but are "taken-for-granted" statements that are known to the experts in the field. The work of the scientist is to present research that allows for as many Type 4 statements as is justifiable. The objective is "to persuade colleagues that they should drop all modalities used in relation to a particular assertion as an established matter of fact" (Latour and Woolgar 1986, p. 80). Ideally, the scientist writes a paper where that claim is subsequently cited as a "fact-like entity" by others.

Claims are modulated through four common linguistic strategies (Fig. 8.9). Probability of something being certain is indicated with words such as "can" "could" "may", or stating that something is "certain", "likely" or "possible". Another way to limit the claim is by generalizing how frequently or in how many circumstances, the claim is applicable. It may be true in "the majority of cases" or "sometimes", for example. The third way to limit the claim is by creating distance from the statement. This is done with phrases such as a result "appears" or "seems" to have occurred for a given reason. Another strategy for creating distance is to take less responsibility for the reason saying "it may be said that it occurred" which does not explicitly state who holds this opinion. Finally, a range of verbs can be used to indicate certainty. The data may "indicate" something, which is less certain than saying they "establish" something. Stating that data "suggest" something is much less certain than they "demonstrate" something.

The linguistic devices for making a claim or reducing its certainty can also be used in combination. A claim may state, "Another possible explanation for the findings described here could be...." This claim uses two forms of probability ("possible" and "could be").

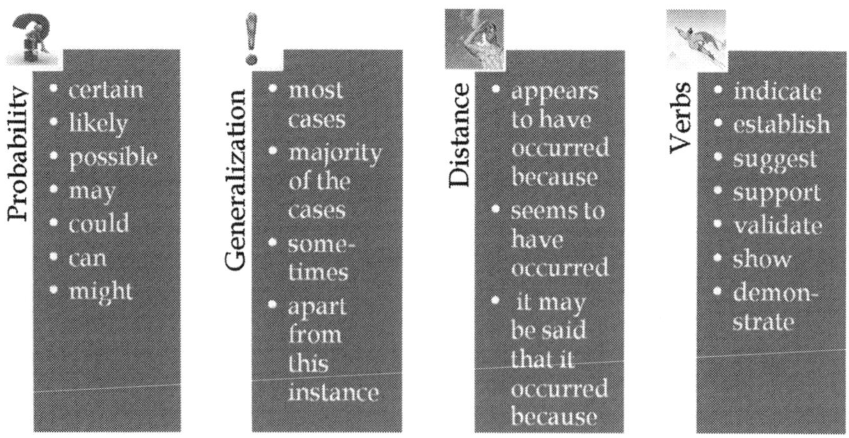

Fig. 8.9 Four strategies to alter the strength of a claim

In another paper, the authors wrote: "Our study indicated that parasite abundance was not correlated with [something], however we did not analyze smaller sizes that are unavailable for [this context]" (Rodríguez-Santiago and Rosales-Casián 2011, p. 201). Two techniques are evident here that limit the generalizability of the finding. The authors use the verb "indicated" which is not especially certain, and they state a limitation of the study (smaller sizes were not analyzed). Together, these techniques reduced the strength of the claim to a level that is acceptable based on the data and the reviewers' comments.

Scientists should aim to make as broad and high a level claim as the data allows, and yet not make such a grand claim that the concept is rejected. Determining the right level of claim is one of the most difficult things for an author to do. Realistically, the claim in the published work is negotiated between the author and the editor. Ultimately, the claim is then determined by the readers and subsequent authors who accept it as stated (a Type 4 claim), or criticize it and lessen its status.

In summary, the Discussion section serves to tell the reader what the data mean and why, what the limitations are in generalizing from the current study, and often it lays out future research that emanates from the current study. It is a nuanced section that makes clear to the reader the contribution of the work and thus changes the knowledge within the field.

8.6 Abstract

The Abstract is the primary "selling feature" of a paper. Apart from the title, it is the section that readers go to in order to determine whether they are interested in looking at the rest of the paper. It is the one section of a paper that all scholars who have access to a database can read, since full access to complete articles is limited to those with an institutional or personal subscription. Thus the Abstract has a two-fold purpose: to entice the reader to read the whole paper, and provide sufficient information that the reader can make an informed decision.

Abstracts tend to fall into two types: they report on Results, or they summarize. Abstracts that focus on results concentrate on the findings themselves and what can be concluded from them. In contrast, summary abstracts put together the essential elements from each section of the paper.

- The Introduction section provides the importance of the area of study, the purpose of the study, and perhaps the research questions or hypothesis.
- The Method section provides essential information about how the study was conducted.
- The Results section is typically handled in one of two ways: either the results are "indicated" or provided in a style called "informative". Examples of indicated results are, "The results of the three experiments are presented" or "The data indicated several variables to be pertinent". The actual results of the

experiments and the pertinent variables are not stated the Abstract. This use of indicated results seems to be less effective as a selling tool for a paper compared to Abstracts that actually inform the reader of the Results. For example, in a study of parasites within fish, the authors included in the Abstract the findings, "The highest values of prevalence and abundance were in *Anisakis* sp. (prevalence = 93.3 %, abundance = 12.4 ± 4.7 ind/host) and *Hysterothylacium* sp. (prevalence = 86.6 %, abundance = 16.5 ± 3.4 ind/host)." Abstracts that state the key results are especially helpful in determining whether or not the paper is of value to the reader.
- The Discussion section provides a statement about the contribution of the paper to the field.

Some disciplines, such as Medicine and Experimental Psychology, may have other more formalized structures for the Abstract section.

Regardless of the field, it is important to consider that the Abstract is the section that a reader sees when searching for scholarly resources. It should be written to emphasize the key importance of the paper based on the study being reported.

8.7 A Second Look at Structures for the Research Paper

The prototypical Introduction, Methods, Results and Discussion structure of research papers has been discussed in detail. Each section serves a purpose that demonstrates the worthiness of the scientist's contribution to the field. The underlying story of the paper goes something like this:

> There is an important area of study and some things that we know about it, but there is something that we don't know. I will fill that gap with this study. I gathered data in a credible manner and this is what the data show. Based on these data, we now know this thing or things that we didn't know before. You can trust these things because of the care and thoughtfulness I put into the study design and the findings' relationship with the relevant research of other scientists. You can use now these ideas as a basis for further conversations in our field.

Not all fields or all journals organize the sections in this way, which is captured in the in the hourglass shape of Fig. 8.1b. A few common alternative structures are shown in Fig. 8.10 below.

Other structures put the Results in the second section, and Methods are reported last. Others mix Results and Discussion together in a reporting a series of studies. Papers that do not have an experimental basis but have a model-building purpose will often replace the Method section with sections that explain theoretical foundations and data gathering assumptions.

It is often most effective for the scientist to examine closely the articles that are typical in the journal which is the intended outlet for a particular research paper. Adopting existing models is a practical method of increasing the likelihood of a paper being accepted.

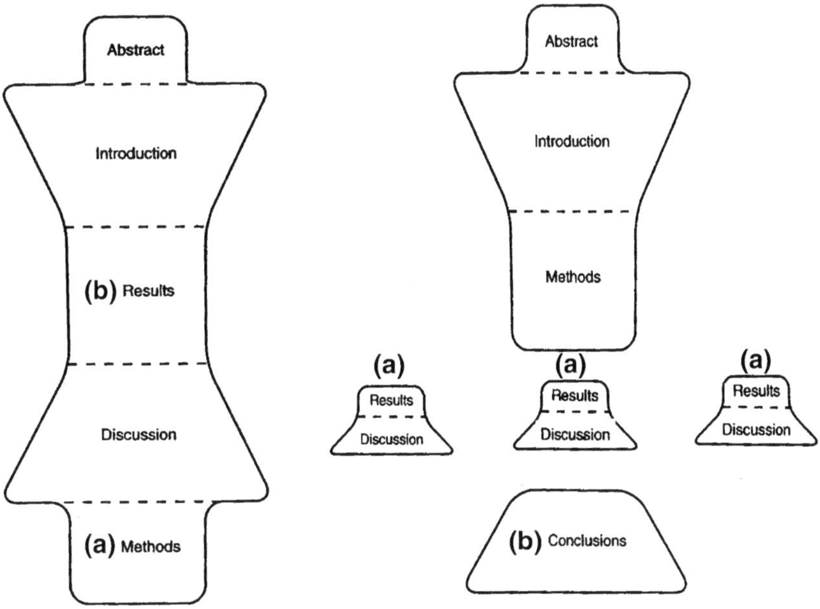

Fig. 8.10 Two common alternative structures for research papers that do not strictly follow the Introduction, Methods, Results and Discussion format (Images from Cargill and O'Connor 2009, p. 11, used with permission)

References

Cargill, M., & O'Connor, P. (2009). *Writing scientific research articles*. West Sussex: Wiley-Blackwell.

Cooper, E. A., Piazza, E. A., & Banks, M. S. (2012). The perceptual basis of common photographic practice. *Journal of Vision, 12*(5), 1–14.

Esquer Mendez, J. L., Hernández Rodríguez, M., & Bückle Ramirez, L. F. (2010). Thermal tolerance and compatibility zones as a tool to establish the optimum culture condition of the halibut *Paralichthys californicus* (Ayres, 1859). *Aquaculture Research, 41*(7), 1015–1021.

Glassman-Deal, H. (2010). *Science research writing: For non-native speakers of English*. London: Imperial College Press.

Latour, B., & Woolgar, S. (1986). *Laboratory life: The social construction of scientific facts*. Princeton: Princeton University Press.

Penrose, A. M., & Katz, S. B. (2001). *Writing in the sciences: Exploring conventions of scientific discourse*. New York: Longman.

Pérez-Flores, M. A., Antonio-Carpio, R. G., Enrique Gómez-Treviño, E., & Ferguson, I. (2011). 3D inversion of control source LIN electromagnetic data. Unpublished manuscript.

Rodríguez-Santiago, M. A., & Rosales-Casián, J. A. (2011). Parasite structure of the Ocean Whitefish *Caulolatilus princeps* from Baja California, México (East Pacific). *Helgoland Marine Research, 65*(2), 197–202.

Swales, J. (1990). *Genre analysis: English in academic and research settings*. Cambridge: Cambridge University Press.

Swales, J. M., & Feak, C. (2004). *Academic writing for graduate students: Essential tasks and skills* (2nd ed.). Ann Arbor: University of Michigan Press.

Chapter 9
Variations in Different Languages and Cultures

Abstract More than 7,000 refereed, academic and scholarly journals publish in languages other than English. Those languages and national contexts often have their own scientific writing conventions. When those conventions are transported into English, they are sometimes considered errors. International variations may include the overall style of writing and the elements found in different sections of research papers. The dominance of English in scientific writing seems to be producing an increasing alignment with English, even in countries that have their own history of publishing scientific research.

Keywords Language • Errors • Proper English • Style • Audience • Reader-responsible • Good writing culture • Portuguese • Spanish • Finnish • French • Chinese •Japanese

Throughout much of this book the scientific research paper has been discussed as though there are only papers published in English. That is not the case (see Chap. 2). There are more than 7,000 refereed, academic and scholarly journals that publish in languages other than English (Lillis and Curry 2010, p. 10). The research indicates that scholarly and scientific publishing in other languages and other nations often have their own writing conventions. The purpose of this chapter is to summarize that research.

Recognizing that different languages can have different science writing conventions is important for three types of readers. First, journal referees and reviewers can benefit by knowing that manuscripts they receive from non-native-English-speaking authors may reflect the conventions that are typical outside the English-speaking world. Thus, variations may seem unusual, but the differences are not necessarily "errors" or the result of carelessness. The variations may simply be a transposition of other language conventions into English, even though the reviewer may deem it necessary that the authors adopt a more conventional English article. Second, teachers who work with scientists

K. Englander, *Writing and Publishing Science Research Papers in English*,
SpringerBriefs in Education, DOI: 10.1007/978-94-007-7714-9_9, © The Author(s) 2014

to support their manuscript writing skills should know that their international scientist-students may have expectations of what a good science article should be, based on reading and writing papers in languages other than English. If these teachers have a better understanding of the different language conventions, they can make explicit the differences and similarities between their students' language and English. Finally, scientists who know languages other than English can use this information to better understand the differences of language conventions that they see in the papers that they read and write.

In this chapter, a summary is presented of the extensive research concerning the contrasts in scholarly research articles published in English and a host of other languages in an array of disciplines. Research concerns the overall style of writing and context, followed by research on different sections of research articles. These contrasts between English and other languages is an active area of research conducted by people who teach or assist non-native-English scholars in their publishing efforts.

9.1 Overall Style of Writing

English has been described as a writer-responsible language (Hinds 1987). The writer is responsible for ensuring that the reader understands the text. In academic writing, texts in English are supposed to be clear in sentence and organizational structure, economical in words, precise in terminology, and inductive in reasoning such that the general precedes the specific, among other common attributes (Bennett 2009). It is expected that the writer will anticipate and eliminate any difficulties the reader may face in understanding. Another language that is considered to be a writer-responsible language is Chinese (Hinds 1987).

In contrast, reader-responsible languages place the onus on the reader to make meaning from the text (Hinds 1987). Reader-responsible texts allow for a much more implicit organization of texts and the reader is considered to be an intelligent being "to whom very little needs to be explained" (Mauranen 1993, cited in Fernández Polo 1999, p. 281). Many European languages such as French, Spanish, Italian, Portuguese, Czech as well as Japanese, are considered reader-responsible languages (Martín-Martín and Burgess 2006). These languages prefer to "leave much to the intelligence and to the imagination" (Fernández Polo 1999, p. 281).

These reader-responsible languages can be so distinct from English that they should be considered a different form of writing. Specific linguistic features can be found in these texts that are not typical of English academic writing (see Table 9.1). These alternate features include very long sentences, complex syntax, figurative language and deliberate ambiguity, among other features. These features are often lauded as elements of good quality writing in the romance languages (Portuguese, Spanish, French, Italian) and in Arabic.

Even within one language tradition (i.e. Portuguese), different disciplines seem to incorporate these non-typical English academic features to different degrees. Based on

Table 9.1 Selected linguistic features of Portuguese, a reader-responsible language, in academic papers, based on Bennett (2010)

Linguistic feature	Explanation/Examples
Magisterial plural	Using "we" within a single-authored paper as form of modesty. E.g. [literal translation] *And, before we advance, let us be permitted to point out…*
Gerunds that allow for ambiguous meaning	For example, in the following excerpt, the gerund 'evitando' ('avoiding') could be interpreted causally ('due to the fact that it avoids'), consequentially ('with the result that it avoids') or merely as additional information about the surgical technique in question ('which avoids'); such ambiguity would not be acceptable in an equivalent medical paper in English. E.g., [literal translation]: *The reconstruction of the anterior cruciate ligament using a graft synthesized* in vitro *by tissue engineering is becoming increasingly common, <u>avoiding</u> the risks of the allograft and the morbidity associated with the harvesting of the autograft.*
Elaborate framing devices	Characteristic embedding of the main clause after an interpersonal framing element. Examples in literal translation include not only the straightforward *it is noted that* or *it should be considered that*, but also highly elaborate frames such as *it would be wrong not to understand that* or *it does not cease to be highly interesting to note that*
Deferred topic	The main idea is deferred to later in the sentence or paragraph, with the effect of creating suspense. E.g. [literal translation] *Founded in 1996 and constituting the only gay periodical, Korpus offers coverage of national gay events,* and *Difficult it becomes, for this reason, to identify meanings and functions.*
Complex syntax	Very long sentences with a great deal of subordination. In the Bennett (2010) study of 408 papers in the humanities and social sciences, typical sentences were 70 words long.
Verbless sentences	A subordinate clause or participle phrase is detached from the main clause and presented as a separate sentence. E.g. [literal translation] *Of which we will assess three. Thus remaining a "mystery" for many orthopedists.*
Multiple negative constructions	A word that already has a negative connotation is accompanied by a negative. While not entirely unacceptable in English, it would be considered convoluted. E.g. [literal translation] *doesn't cease to be/show/cause concern…*or *not rarely,* or *this does not mean that they do not continue to be….*
Poetic, figurative or high-flown diction	Vocabulary used to characterize entities would be considered hyperbolic or ironic in English. For example, a university is referred to as *an alma mater, whose body illuminates time with the lights of knowledge.*
Archetypal singular	A singular noun is used with the definite article to refer to a collective and in so doing creates a generalization. E.g. [literal translation] *the adolescent prefers* or *the Trade Unionist believes…* English academic prose would expect the plural form for collective entities, as in "adolescents prefer".

Bennett's (2010) survey of 408 Portuguese published papers in Medicine, Economics and Engineering indicates they rarely deviated from the English model. Law, Education and Psychology diverged somewhat from the English model. The most elaborate styles were found in Archaeology, Architecture, Sociology, History and Philosophy, and direct translation of these papers would not by readily acceptable in the English model. The language of these texts is "typically ornate, erudite and complex, with a great deal of redundancy, which makes it sound rather pompous, longwinded, and old-fashioned to the English ear" (Bennett 2010, p. 30). Only 4 % of the texts contained none of the linguistic features of a reader-responsible language.

9.2 Introduction Sections and Moves

Introduction sections have a typical structure in English-language research papers, as was discussed extensively in Chap. 8. Briefly, based on the research of John Swales (1990, 2004), there are three "moves" or elements that serve to create an intellectual space for the particular research paper. Move One establishes the research territory, including reviewing the existing research which may be insufficient or misguided in dealing with a particular scientific phenomenon. Move Two establishes a niche or a gap in the literature. Move Three fills that gap by stating the objective of purpose of the current study. Of course, not every paper will neatly perform the three moves and present them in a sequential fashion, but the vast majority do.

Swales (2002) has said that the tripartite move structure may be particularly appropriate to the English-speaking countries with large research investments. This structure works well because this context is a "big world… with big journals". The competition to create an intellectual space may be strong in a nation that hosts many researchers and many prestigious journals. There is a need to create a research space for the scientist's work in relation to the rest of the field (Swales 2002).

Researchers have examined Introductions published in scholarly journals in other languages and other national contexts to identify their adherence to or variation from the three-move structure. Their findings indicate:

- Japanese scientists tend to avoid an overt Move Two where they would state an obvious gap in the literature (Gosden 1995).
- In Spanish-language abstracts, a clear statement of the gap appears in only 15 % of the papers (Martín 2003).
- In Malay, Thai and Arabic, typically, there is not a "gap" statement (Loi 2010).
- In Chinese psychology papers, the authors may point out where there is a lack of research on a topic, but they do not criticize specific researchers or their published studies, which is a common practice in the English papers (Loi 2010).
- Also, Chinese papers in Psychology usually do not state the implications of the findings in the Introduction, while comparable English papers do, but the overall Introduction structure in Chinese papers is consistent with English models (Loi 2010).

- Japanese scientists avoid confrontational ways of framing the need or contribution of their papers (Gosden 1995).
- Korean (Cho 2004) and Sudanese (ElMalik and Nesi 2008) scholars avoid being perceived as bragging or boosting their work.
- Hong Kong Chinese scholars seem reticent to be critical of other authors (Flowerdew 1999a), and they may deliberately not cite relevant texts that they consider to be "deficient or incomplete" (Hyland and Salager-Meyer 2008, p. 317).

Two elements that are typical of English-language Introductions are not typical in other contexts: confrontations with other researchers and explicit statements of a gap in the existing literature.

The literature review section of the Introduction is sometimes underdeveloped by scholars based in developing countries. Access to thorough and up-to-date bibliographic materials may be limited by lack of monies for books or journal subscriptions. Scholars such as those in Sri Lanka may adopt a strategy of "the less said the better" about the existing research. By saying less, they are less likely to create an "incomplete literature review that would attract unnecessary attention to their limitations, generate criticism from referees, and jeopardize their chances of getting published" (Canagarajah 2006, p. 599).

9.3 Discussion Section

A principal element of the Discussion is making claims about the findings, as discussed in Chap. 8. Determining the appropriate level of claim based on the scientific evidence is crucial to a paper's acceptance. International authors report that the Discussion section is often the most difficult to write (Flowerdew 1999b; Swales 2002; Moreno et al. 2012), and this may be true in part because the conventions differ from English language science writing style.

Research on the differences has been conducted for a number of languages and cultures. For example, Hong Kong Chinese scholars can be "too assertive" in their claims compared to what is accepted in English language papers (Flowerdew 1999b, p. 256). Mexican scientists can also attribute greater importance to their work than is typical in English and so they are accused by journal reviewers of hyperbole (Englander 2009). Other researchers downplay their contributions, such as Spanish social scientists and biomedical scientists (Burgess 2002; Pérez-Llantada 2010) and Malaysian scientists (Ahmad 1997).

If we compare the presence of hedges, boosters or attitude markers (see Chap. 7) in Spanish published papers and English ones, these linguistic devices are much more common in English (Hyland 2000; Martín-Martín 2008). In Spanish, hedges, boosts and "attitude markers" were used once every 80 words compared to once every 28 words in English.

There is variation in the frequency of hedges and boosters in different disciplines, especially in the Introduction and Discussion/Conclusion sections of published papers. The lack of "boosterism" has been explained as the consequence of being part of relatively small research communities where the scientists rely on the authors they are citing for other collaborative and funding activities, so their statements of claim must be carefully respectful of their national colleagues (Burgess 2002). In larger academic communities, such as the U.S., such caution may be less necessary.

Authors' statement or re-statement of their contribution which forms part of the Discussion or Conclusion is also done more visibly in English language texts than in, for instance, Italian papers (Molino 2010). The comparison can be seen in two Linguistics papers: an English-language scholar wrote: "In this paper, I have argued that…" whereas an Italian linguist wrote: "The idea, as has been said, was…." (translated from Italian; Molino 2010, p. 97). The scholars' presence is foregrounded in the English paper and backgrounded in the Italian one.

Another distinguishing feature of Spanish biomedical papers is the authors' manner of arguing for their interpretations: Whether they are publishing in Spanish or in English, they construct "a more elaborate and digressive discourse" than is typical of English authors (Pérez-Llantada 2010, p. 63). It is interesting to note, however, that when Spanish biomedicine scholars publish in English, the rhetorical style is closer to English papers than when they are publishing in Spanish. This seems to indicate that they adjust their writing to more closely mimic the English academic style and lessen the Spanish rhetorical features when they address an English-language audience.

Differences can be seen not only in *how* the Discussion claims are written, but *which kinds* of claims are included. A study that compared articles in prestigious medical journals written by British and Sudanese researchers found that the papers were similar in most respects. All the authors from both countries included statements to highlight the overall outcome and explain specific outcomes of their research (ElMalik and Nesi 2008). However, a third of the Sudanese writers did not include statements that indicated the significance of the outcome. The researchers note that stating the significance of a finding is an important persuasive element that can give way to polemic, particularly if the writer lacks sophisticated linguistic skill (ElMalik and Nesi 2008). An author's uncertainty about managing a delicate, persuasive element of a manuscript can lead the author to simply avoid it. This sentiment is captured in the title of another investigation into Spanish scientists' publishing experience, "You Don't Say What You Know, Only What You Can" (Pérez-Llantada et al. 2011). As a consequence, the role of the research in the larger conversation of the field is not provided to the reader.

9.4 Abstract Section

Less research has focused specifically on Abstracts. However a few differences between Swedish and English, and Turkish and English authors have been documented. Abstracts published in Swedish are not very consistent in their presentation of content. This may be because the Abstract is less important for entry into and promotion of the paper in this smaller academic environment (Melander et al. 1997).

The community of scholars is more likely known to each other, and so the Abstract is not as necessary to entice readers to the paper's content. Unlike Abstracts in English, Abstracts in Turkish science articles rarely used personal pronouns of "I" or "we" and they were more likely to use the passive voice than active voice. According to Cakir (2011), the passive voice has the effect of creating an objective stance of the authors (Cakir 2011).

9.5 Context

Before leaving this chapter that describes specific differences in the discourse in different languages, it is important to point out that context itself—the intended audience and the journal readership—can play a crucial role in crafting a research paper. As an example, consider the case of a Sri Lankan social scientist who wrote two papers in English, one for a predominantly local audience as part of an edited volume, and the other for an international readership (Canagarajah 2006). Differences were found in the rhetoric of the two papers. The paper for the national audience represented his respect for local ethos and the humility that is expected within the national scholarly tradition. In contrast, the paper for the international audience was more argumentative and linear in development. This is evidence that he styled his paper to account for his English-speaking audiences, apart from the linguistic conventions of his first language, Tamil. Thus, we should be careful not to attribute differences in manuscripts written by multilingual writers solely to the interaction between the first language and English. There may be other elements in the context that the author is addressing in the rhetoric of the text.

9.6 Summary and Conclusions

Different languages and cultures present scholarly research articles to some extent differently than the English/Western version that is widely expected in English-language journals today. The rhetoric of some languages may demand that the writer has the responsibility to be sure that the reader understands the content, whereas others expect the readers to make sense of the text. Introductions, Discussions, and Abstracts may be structured in a different manner and use different linguistic or rhetorical styles than English. These may be linguistic differences that reflect academic communities that are smaller and more interdependent than English-language international ones. Studies of non-native-English speaking academics find that when the non-English elements are carried into English-language texts, the authors are admonished by journal reviewers for submitting manuscripts that are awkward, poor, or linguistically insufficient (Hanauer and Englander 2013). Despite whatever linguistic and rhetorical academic writing traditions an international scholar may value, it seems that they are often expected to produce manuscripts which demonstrate "a virtual effacement of the L1 (i.e. first language or mother tongue) abilities and predispositions" (Byrnes 2002, p. 42).

Based on the research that compares papers published in English by native speakers and those in other languages over time, there is increasing alignment with English. In other words, the rhetorical and organizational styles that scholars produce when writing in languages other than English are becoming closer to the styles and organizational structures of English (Bennett 2010; Martín-Martín and Burgess 2006). This alignment with English is more common in disciplines of the natural sciences than in the social sciences and humanities. This increasing alignment can be accounted for due to two reasons. First, the prestige and domination of English language publishing may be further effacing native language predispositions and thus influencing how scholars write in their native language. Second, cultures and languages that do not have a long history of Western-style scientific research—that is, *not* Germany, France, and Switzerland—emulate English writing models as the basis for writing science in their own language.

References

Ahmad, U. K. (1997). Research article introduction in Malay: Rhetoric in an emerging research community. In A. Duszak (Ed.), *Culture and styles in academic discourse* (pp. 273–303). Berlin: Mouton de Gruyter.

Bennett, K. (2009). Academic style manuals: A survey. *English for Academic Purposes, 8*(1), 43–54.

Bennett, K. (2010). Academic discourse in Portugal: A whole different ballgame? *English for Academic Purposes, 9*(1), 21–32.

Burgess, S. (2002). Packed houses and intimate gatherings: Audience and rhetorical structure. In J. Flowerdew (Ed.), *Academic discourse* (pp. 196–215). Harlow: Longman.

Byrnes, H. (2002). Toward academic-level foreign language abilities: Reconsidering foundational assumptions, expanding pedagogical options. In B. L. Leaver & B. Shektman (Eds.), *Developing professional-level language proficiency* (pp. 34–58). Cambridge: Cambridge University Press.

Cakir, H. (2011). Forms of constructing scientific knowledge and encoding writer's stance in Turkish and English research article abstracts. Unpublished dissertation, Dokuz Eylül University, Turkey.

Canagarajah, A. S. (2006). Toward a writing pedagogy of shuttling between languages: Learning from multilingual writers. *College English, 68*(6), 589–605.

Cho, S. (2004). Challenges of entering discourse communities through publishing in English. *Journal of Language, Identity and Education, 3*, 47–72.

ElMalik, A. T., & Nesi, H. (2008). Publishing research in a second language: The case of Sudanese contributors to international medical journals. *English for Academic Purposes, 7*, 87–96.

Englander, K. (2009). Transformation of the identities of nonnative English speaking scientists. *Journal of Language, Identity and Education, 8*, 35–53.

Fernández Polo, F. J. (1999). *Traducción y rétorica contrastiva: A propósito de la traducción de textos de divulgación científica del inglés al espanol*. Santiago de Compostela, Spain: Universidad de Santiago de Compostela Servicio de Publicaciones.

Flowerdew, J. (1999a). Writing for scholarly publication in English: The case of Hong Kong. *Journal of Second Language Writing, 8*(2), 123–145.

Flowerdew, J. (1999b). Problems in writing for scholarly publication in English: The case of Hong Kong. *Journal of Second Language Writing, 8*(3), 243–264.
Gosden, H. (1995). Success in research article writing and revision: A social-constructionist perspective. *English for Specific Purposes, 14*, 37–57.
Hanauer, D. I., & Englander, K. (2013). *Writing science in a second language*. West Lafayette: Parlor Press.
Hinds, J. (1987). Reader versus writer responsibility: A new typology. In U. Connor & R. Kaplan (Eds.), *Writing across languages* (pp. 141–152). Reading: Addison Wesley.
Hyland, K. (2000). Hedges, boosters, and lexical invisibility: Noticing modifiers in academic texts. *Language Awareness, 9*(4), 179–197.
Hyland, K., & Salager-Meyer, F. (2008). Scientific writing. *Annual Review of Information Science and Technology, 42*(1), 297–338.
Lillis, T., & Curry, M. J. (2010). *Academic writing in a global context: The politics and practices of publishing in English*. London: Routledge.
Loi, C. K. (2010). Research article introductions in Chinese and English: A comparative genre-based study. *English for Academic Purposes, 9*, 267–279.
Martín, M. P. (2003). A genre analysis of English and Spanish research paper abstracts in experimental social sciences. *English for Specific Purposes, 22*, 25–43.
Martín-Martín, P. (2008). The mitigation of scientific claims in research papers: A comparative study. *International Journal of English Studies, 8*(2), 133–152.
Martín-Martín, P., & Burgess, S. (2006). Reader and writer responsibility in abstracts in Spanish social sciences journals. *Explorations in Specialized Genres, 35*, 43–57.
Melander, B., Swales, J. M., & Frederickson, K. M. (1997). Journal abstracts from three academic fields in the United States and Sweden: National or disciplinary proclivities? In E. Duszka (Ed.), *Culture and styles of academic discourse* (pp. 251–272). Berlin: Mouton de Gruyter.
Molino, A. (2010). Personal and impersonal authorial references: A contrastive study of English and Italian Linguistics research articles. *English for Academic Purposes, 9*(1), 86–101.
Moreno, A. I., Rey-Rocha, J., Burgess, S., López-Navarro, I., & Sachdev, I. (2012). Spanish researchers' perceived difficulty writing research articles for English-medium journals. *Ibérica, 24*, 157–184.
Pérez-Llantada, C. (2010). The discourse functions of metadiscourse in published academic writing: Issues of culture and language. *Nordic Journal of English Studies, 9*(2), 41–68.
Pérez-Llantada, C., Plo, R., & Ferguson, G. R. (2011). "You don't say what you know, only what you can": The perceptions and practices of senior Spanish academics regarding research dissemination in English. *English for Specific Purposes, 30*(1), 18–30.
Swales. (1990). *Genre analysis: English in academic and research settings*. Cambridge: Cambridge University Press.
Swales, J. (2002). On models in applied discourse analysis. In C. N. Candlin (Ed.), *Research and practice in professional discourse* (pp. 61–77). Kowloon: City University of Hong Kong Press.
Swales, J. M. (2004). *Research genres: Explorations and applications*. Cambridge: Cambridge University Press.

Part III
The People

Doing scientific work and being a scientist is a social activity involving many people in many different ways. Scientific publishing, too, takes place in a social context filled with people—from graduate students learning to become scientists, to emeritus professors who edit international scholars' papers for publication as a service to the field. Thus, understanding the social roles of people in networks that promote scientific writing competence or that create barriers for participation can be vitally important.

Part III examines the research about people and the roles they play in relation to scientific writing and publishing. The research comes from fields such as sociology, education, applied linguistics, and others. Much of this research is centered on the idea of participating in a "community of practice"; this is a term used in applied linguistics. A community of practice describes a group of people who come together to perform mutually important and understood activities. It captures the idea that people participate in a given community, and it indicates an approach in which people "learn" to become something (a butcher, a particle physicist, a gang member, and so on) by learning from others in the "community" who "practice" this same kind of identity.

In this part, we will explore the different roles that people have within the science community and discuss how participation in these networks can promote successful experiences in writing for scientific publication.

Chapter 10
Graduate Students Becoming Scientists

Abstract Students must learn to "know" science and to "be" scientists during their academic training. Apprenticeship-like relationships, also called enculturation, permit students to learn the academic and social skills that science requires. Sometimes excluded from helpful mentoring relationships are international students, women and minorities. Good relationships in graduate settings can lead to scientific collaborations that are both formal and informal long after graduation. The relations forged in graduate school are shown to be crucial to scientific success.

Keywords Advisors • Supervisors • Academic relationships • Mentoring • Women • Minorities • Enculturation • Apprenticeship • Mentors

In graduate school, students transform from not only "knowing" science but "being" scientists. This is typically a process of enculturation where students' knowledge of the discipline broadens to include the "ways of thinking, believing, valuing, and using various symbols, tools, and objects" that established members of the discipline do (Gee 2011, p. 29). In applied linguistics, this is called participating in the big "d" Discourse of the field. For this to occur, increasingly close interaction over a long time with established members is necessary.

Graduate students report that their relationships with their supervisors are crucial to their satisfaction and eventual success (DeWelde and Laursen 2008; Hanauer and Englander 2013). "Outstanding advisers challenge their students, set high expectations and standards, generously share their expertise, and individually tailor their students' educational experiences" (Golde et al. 2006, cited in DeWelde and Laurson 2008, p. 59). Unfortunately, such a relationship is not as common as desired. A large-scale study with 1,000 graduate students found that a little over half (56 %) were "satisfied" or "very satisfied" with their supervisory relationship (Ortolani 1998, cited in Nakamura et al. 2009, p. 2).

In contrast, an advisor's lack of interest in a student's work or unavailability for consultation in the research or publication process has caused more than one

K. Englander, *Writing and Publishing Science Research Papers in English*, SpringerBriefs in Education, DOI: 10.1007/978-94-007-7714-9_10,

student to leave science altogether (DeWelde and Laursen 2008; Holmes and O'Connell 2003; Ivanic 1998). In a most extreme case, 20 years ago a Nobel laureate's star graduate student in Harvard's chemistry department committed suicide; the note he left behind stated, "this event could have been avoided… professors here have too much power over their students" (Hall 1998, cited in Nakamura et al. 2009, p. 2). While such overwhelming helplessness is horrific, lesser feelings of alienation towards supervisors seem to be still too common.

Students generally desire advisors who provide opportunities to write up their work, present at conferences, participate in grant writing, and take intellectual risks. These advisors set the foundation for students to feel connected to their discipline and perceive themselves as contributors to the knowledge of their field.

International graduate students often seek to study in the "developed" world in order to gain what they perceive as the best possible education. These educational opportunities are more likely to lead to later collaboration in professional life depending on the "mode of study" and the "mentoring style" of their graduate education. Mode of study refers to whether the graduate degree is earned by research and coursework combined, only research, or only coursework. Mentoring style is defined by frequency and intensity of interaction, degree of supervision and amount of time spent on research. Research by Ynalvez and Shrum (2009) indicate that where the mode of study is by research only, a strong bond is developed between student and supervisor that is more likely to carry on beyond graduation. Such a mode of study is likely to co-exist with a mentoring style that is frequent, intense and involving personal interaction. Together, this creates a learning environment with "a sense of community" and "the development of durable and strong professional ties" (Ynalvez and Shrum 2009, p. 892). The only-coursework mode does not seem to foster close alliances between student and professor.

Research into the experience of women and minorities (African-Americans, Hispanic-Americans) in graduate study in the United States may explain some of their lesser representation in the sciences and engineering. Women disproportionately report *not* receiving career advice and experiencing dissatisfaction with their advisors (DeWelde and Laursen 2008). Women and minorities are more likely to not complete their doctoral education, and if they do, more likely to work outside of academia. It is important to note, however, that this may be largely a North American finding. The former East Germany (Hanson et al. 2004) and other nations of Eastern Europe count a higher proportion of women in science (39–43 %, UNESCO 2006) than does the United States (27 %). Parity between men and women in science is most often found in Latin America: specifically Argentina, Brazil and Venezuela (UNESCO 2006).

Scientific writing competence has been shown to be developed by distinct activities through higher education, particularly for those who study in countries where English is not the dominant language. The trajectory shown in Fig. 10.1 indicates that during bachelor and masters study, students become immersed in reading scientific papers and writing is typically in a classroom, group context. When students begin writing their masters thesis and move into doctoral study, the writing becomes individualized, ideally in close collaboration with senior

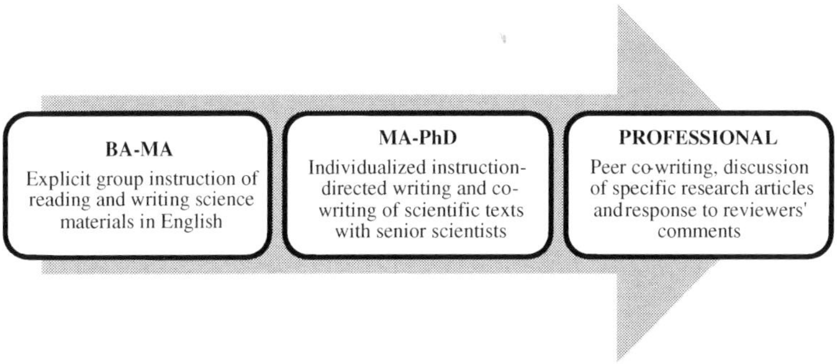

Fig. 10.1 Important educational activities that develop scientific writing skills during the educational trajectory (Reprinted from Hanauer and Englander 2013, p. 125, used by permission)

scientists. Honing the practice of writing research papers in English continues into professional life.

Learning the social interactions around publishing is sometimes an under-recognized activity in graduate school. While doctoral programs tend to make students aware of the central role of publishing, the relationships that can facilitate publishing are sometimes neglected, and are rarely overt elements of graduate study. Students benefit greatly by "establishing close relationships with influential members of the discipline, including accomplished researchers, editorial board members and editors of journals" (Long et al. 2009, p. 236).

Successful scientists report that their earliest papers were written in close consultation with their advisors at the graduate level (Hanauer and Englander 2013). The content of the interactions can take many forms: determining a research question, planning and implementing the research, data analysis, and situating the findings in the larger literature. Such discussions between a graduate student and supervisor are central, and they underpin research writing. Translating these discussions into suitable text usually requires multiple iterations, including discussion of the structure of research papers, distinguishing the content of Introductions versus Discussions, and Results versus Discussions.

Graduate students also learn from advisors that the typical patterns of a research article can be somewhat flexible. For example, a recent study of how scientists promote scientific inquiry illustrates how a professor helped his student to see that an article doesn't have to strictly follow a literature-review-then-results format. In this case, a student worried that his three-stage experiment did not fit the traditional research paper structure, so the professor responded by saying:

> [It] doesn't have to be that way. So when you finish [one] section you have got a literature review, you've got your method, you've got some results. Then, there's Stage 2, with a lit review, method and results. Then there's Stage 3, lit review, method, results. You don't have to have all the background literature in one section of the paper. (Uzuner-Smith and Englander, under review).

Close interaction with a supervisor is found to be crucial to students' success in learning to write journal articles. The sooner that this process begins, the better.

The more that students who have the opportunity while undergraduates to see real research in action, the more likely they are to complete the trajectory to engage in a scientific career (Hanauer et al. 2012). Real research in action includes the creation of research papers. Establishing a good relationship between the supervisor and the graduate student has benefits that long exceed the time spent in graduate school. Such relationships are to be encouraged.

References

DeWelde, K., & Laursen, S. (2008). The "ideal type" advisor: How advisors help STEM graduate students find their 'scientific feet'. *The Open Education Journal, 1*, 49–61.

Gee, J. P. (2011). *An introduction to discourse analysis* (3rd ed.). London: Routledge.

Hanauer, D. I., Frederick, J., Fotinakes, B., & Strobel, S. (2012). Linguistic analysis of project ownership for undergraduate research experiences. *Life Sciences Education, 11*, 378–385.

Hanauer, D. I., & Englander, K. (2013). *Writing science in a second language*. West Lafayette: Parlor Press.

Hanson, S., Fuchs, S., Aisenbrey, S., & Kravets, N. (2004). Attitudes toward gender, work, and family among female and male scientists in Germany and the United States. *Journal of Women and Minorities in Science and Engineering, 10*, 99–130.

Holmes, M. A., & O'Connell, S. (2003). *Where are the women geoscience professors?*. Washington, D.C.: National Science Foundation/Association for Women Geoscientists.

Ivanic, R. (1998). *Writing and identity: The discoursal construction of identity in academic writing*. Amsterdam: Benjamins.

Long, R., Crawford, A., White, M., & Davis, K. (2009). Determinants of faculty research productivity in information systems. *Scientometrics, 78*, 231–260.

Nakamura, J., Shernoff, D., & Hooker, C. (2009). *Good mentoring: Fostering excellent practice in higher education*. San Francisco: Jossey-Bass.

UNESCO. (2006). Women in science: Under-represented and under-measured. *UIS Bulletin on Science and Technology Statistics, 3*, 1–7.

Uzuner-Smith, S., & Englander, K. (under review). Investigating scientific inquiry as manifested in talk.

Ynalvez, M. A., & Shrum, W. (2009). International graduate students training and scientific collaboration. *International Sociology, 24*, 870–901.

Chapter 11
Novice Scientists and Expert Scientists

Abstract Novice scientists usually aim to establish their presence in the field and produce the requisite publications that will lead to favorable tenure decisions. Rarely are responsibilities made explicit in a contract or an institutional regulation, thus novices must learn the local university and departmental culture by absorption. The pedigree a scientist brings to a university position can help in establishing a presence, but pedigree does not assure such a presence or publishing success. Establishing networks, seeking out a mentor and forming informal relationships can promote a novice's integration into the discipline.

Keywords Tenure • Implicit learning • Enculturation • Academic affiliation • Explicit learning • Academic appointments • Predicting scientific success

Novice scientists are considered those with fewer than 6 years beyond their doctorate. Typically, this 6-year period accounts for the time between completion of the doctorate degree and a university tenure decision. This period is crucial for establishing a presence in the field and producing the requisite publications that lead to a favorable tenure decision. Thus understanding the challenges and opportunities that novice scientists face can help to assure their continued participation in the scientific community.

One means of attaining an entry into a professorship is to graduate from a prestigious school. An advantage of attending a high-status university, and particularly a high-status university in the Anglophone world, is the "pedigree" it provides that "has the effect of opening doors and creating opportunities" for graduates. These young researchers may then benefit from "homosocial reproduction" (D'Aveni 1996, cited in Long et al. 2009, p. 236): they are hired as incoming junior faculty by senior faculty who are graduated from those same institutions.

Graduation from a high prestige school is not always a significant predictor of publication success. Novice scholars seem to be more greatly influenced by the institution she or he is affiliated with than the school they graduated from.

K. Englander, *Writing and Publishing Science Research Papers in English*, SpringerBriefs in Education, DOI: 10.1007/978-94-007-7714-9_11, © The Author(s) 2014

This phenomenon is borne out by two studies. In one study, authorship was studied in the top journals in chemistry, economics, philosophy and sociology over a 25-year period (McNamee and Willis 1994). The researchers found that publication rates were more favorable for scientists based at "mid-level" universities. Often, these are people who attended first-tier universities but due to market forces, were unable to obtain a post at a similar institution. These individuals seem to be especially motivated to display knowledge contributions to their scientific discipline. Long et al.'s (2009) study of the information systems discipline also found that "academic affiliation is a significant determinant of productivity in terms of quantity (as measured by publication counts) and quality (as measured by citation counts).... Status of a faculty member's academic origin is not a significant determinant of research productivity" (p. 231).

The fact that affiliation is a greater determiner of productivity than training indicates that the institutional environment that a novice scholar enters has a great impact.

The president emeritus of Stanford University has commented on the often unclear road that a novice faculty member must navigate to achieve tenured status. The knowledge about functioning effectively as a scientist-scholar is "learned by absorption" (Kennedy, cited in Nakamura et al. 2009, p. 1). He continues by saying:

> The faculty member's understanding of his or her academic responsibilities is not prescribed by contract or institutional rule; in this respect it is unlike the understanding of duty one would have as a soldier or a mid-level executive in a large corporation. It is, rather, part of an inherited culture (Kennedy, cited in Nakamura et al. 2009, p. 1).

Scientific anthropologists Berkenkotter and Huckin (1995) write that this cultural learning "is 'picked up' in the local milieu of the culture rather than being explicitly taught" (p. 11). Learning the inherited culture is a key requirement for success in the academic-scientific world. The culture is not a monolithic one; rather, it varies by discipline, by institution, and often by department.

To learn the culture, participation in the activities of the discipline is crucial. Attending conferences, giving poster presentations, and publishing in conference proceedings can be important entries into the field. For instance, a recent editorial in the *Journal of the American Medical Association* encouraged medical novices to participate in such activities "as it can play a beneficial role in helping become better journal authors" (Stewart et al. 2013, p. 100). Further, such participation allows novices to "build contacts with peers and colleagues" (Stewart et al. 2013, p. 100).

Universities can foster novice scientists' productivity in a number of ways: through mentorship, funding resources, and collegial culture. Novices who seek out such institutional environments are likely to succeed.

References

Berkenkotter, C., & Huckin, T. (1995). *Genre knowledge in disciplinary communication: Cognition, culture, power*. Mahwah: Erlbaum.

Long, R., Crawford, A., White, M., & Davis, K. (2009). Determinants of faculty research productivity in information systems. *Scientometrics, 78*, 231–260.

McNamee, S. J., & Willis, C. (1994). Stratificaiton in science: A comparison of publication patterns in four disciplines. *Science Communication, 15*, 396–416.

Nakamura, J., Shernoff, D., & Hooker, C. (2009). *Good mentoring: Fostering excellent practice in higher education*. San Francisco: Jossey-Bass.

Stewart, M., Chandra, R., Chiu, A., Hanna, E., Kennedy, D., Kraus, D., et al. (2013). The value of resident presentations at scientific meetings. *JAMA Otolaryngol Head Neck Surgery, 139*(1), 100.

Chapter 12
English-Speaking Scientists and Multilingual Scientists

Abstract Scientists who do not speak English as their mother tongue report difficulties in bringing their manuscripts to successful publication in English-language journals. Writing is found to be 21 % more difficult, on average, than writing a scientific paper in their native language. Barriers include sentence-level elements of vocabulary and syntax; however, issues of manuscript organization, methodological criticisms, and devaluation of non-English bibliographic references are also reported. The increased variation in the English that is publishable indicates that the language is growing as a lingua franca.

Keywords Non-native speaker • Native speaker • ESL • Difficulty writing • Anglophone • Periphery • Developing countries • English as an additional language • Lingua franca

Scientists who live and work in non-Anglophone countries tend to have greater difficulty getting their scientific work published in English-language journals than people who speak English as their native language and who are based in English-speaking countries (See Chap. 9 for more detail).

The difficulty can be quantified. In one study, Spanish-speaking scientists in Mexico were asked how difficult it was to write a scientific article in their mother tongue (Hanauer and Englander 2013). They were asked to rate the difficulty on a scale of 1 to 7 where 1 was 'very easy' and 7 was 'very difficult'. Then they were asked how difficult it was to write a scientific article in English on an identical scale. The difference between the two ratings is the difficulty of writing in a second language. Writing in English was shown to be 24 % more difficult (Fig. 12.1). When the same method was used to investigate Spanish-speaking scientists in Spain, the result showed that writing in English was 21 % more difficult (Moreno et al. 2012). It should be noted that these results are the *means* of self-reported data. Not *all* scientists report that writing in English is more difficult than in their mother tongue (Hanauer and Englander 2013; Matsumoto 1995). According to

K. Englander, *Writing and Publishing Science Research Papers in English*, SpringerBriefs in Education, DOI: 10.1007/978-94-007-7714-9_12,

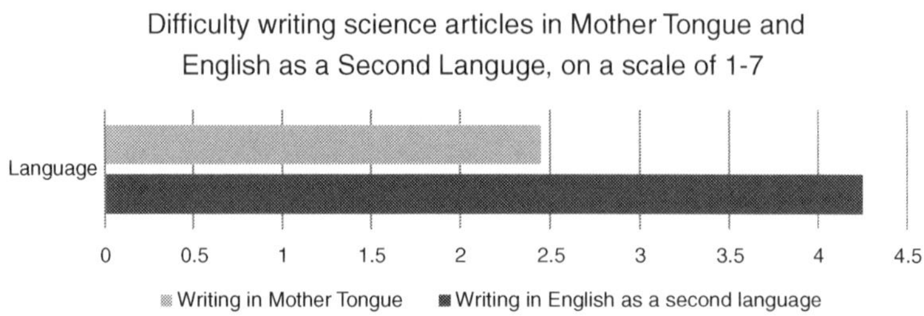

Fig. 12.1 Means (Writing in Mother Tongue 2.45; Writing in English as a Second Language 4.25) on a scale from 1 to 7 of self-perception ratings on the question of difficulty of writing in mother tongue (Spanish) and second language (English) scientific writing (n = 148) indicating an increased difficulty of 24 % (based on Hanauer and Englander 2011)

Moreno et al. (2012), that difficulty typically does not decrease until the scientist has at least ten years of writing experience. Nonetheless, on average, there is a significant increase in difficulty in writing in the second language over writing in one's mother tongue (Uzuner 2008).

For non-native speakers of English, all the sections of a research article are more difficult to write in English than in their native language: Introduction, Method, Results, Discussion and Conclusion. In particular, the Discussion section of articles is reported as the most difficult section to write (Gosden 1996; Moreno et al. 2012).

Non-native speakers of English tend to report that sentence-level elements are often the most difficult to master. Choosing the right vocabulary, using the accepted prepositions in multi-word phrases ('as seen in', 'based on', 'on the other hand'), constructing sentences correctly and assuring that subjects and verbs agree are some of the elements that are time-consuming to produce well when one lacks native-speaker familiarity with the language. As noted by Huang (2010) the scholars' lack of English proficiency subjects them to great pressure and contributes to their perception of being disadvantaged.

The difficulty in writing in a second language supports the claim by Ammon (2001). He writes that the non-native speakers of English bear a greater cost to produce "linguistically adequate texts" than do native-English speakers (p. 7). Ammon points out:

> It seems almost self-evident that the native speakers of the prevalent scientific language [English] have less difficulty using it… and, therefore, have advantages over [non-native-English speakers]…. It is easier for them to produce utterances and texts in line with the existing native-speaker norms (Ammon 2001, p. 7).

The expectation that all scientists write "like a native speaker" seems to be changing somewhat, and that is a welcome change for non-native speakers. English is increasingly considered a *lingua franca* or the common working language of science, and thus its form may carry traces of writers' mother tongue. For instance, analysis of some 500 scientific papers published in English by Spanish scientists shows that the papers retain features of Spanish style of syntax and argumentation (Pérez-Llantada 2012). There is clear evidence that "the Spanish scholars retain part of their culture-specific intellectual style when they write in English as an additional language" (Pérez-Llantada 2012, p. 104). This result suggests that strict adherence to English-language norms of scientific style is not required to be accepted for publication.

Another study analyzed the language used in "Best Paper" award winners in the journal *IEEE Transactions* (Rozycki and Johnson 2013). These papers showed many instances of non-standard English grammar, suggesting that there is some degree of acceptance for unusual English usage, especially by the mostly non-native speakers of English who wrote the award-winning papers. This is a welcome recognition on the part of journal reviewers and editors of the international scope of science. Extended discussion of the role of reviewers is presented in Chap. 13.

Examination of reviewer comments to international scholars indicates that the criticism is largely based on an evaluation of what knowledge is considered to be of interest to a global academic community. A psychology paper written by a Spanish scholar was considered by reviewers to be too limited since it was based on data from people in Spain (Lillis and Curry 2010). The reviewers questioned why readers of a U.S.-based journal would be interested in Spanish data. The research was considered too local to be generalizable. This so-called local data is more often accepted when it is presented as either a contrast to or a comparison with data from an Anglophone nation. On the other hand, English-language journal reviewers seem to have a presumption that data collected in the Anglophone context is universally applicable.

As a result, international researchers can feel they are caught in a double bind: "If they foreground the local, they may be accused of being parochial [and not of universal importance]; if they background the local, they may be denied claim to universal relevance" (Lillis and Curry 2010, p. 154). This experience of international scientists being rebuked for attempting to position their work as being of importance in the Anglophone world has been documented in the natural sciences as well as the social sciences (Aydlini and Mathews 2000; Englander 2009; Hwang 2005).

Gaining acceptance in English-language journals is sometimes facilitated by collaborating in research projects with researchers in the Anglophone countries and in the developed world. Such collaboration between researchers in the Anglophone countries and other parts of the world is increasing (Pérez-Llantada 2012). Non-native-English-speaking scientists report that these collaborations can ease the journal review process because the authors are less likely to be criticized for idiosyncratic use of English or language errors (Hanauer and Englander 2013).

In addition, collaboration can give international researchers access to developed world resources that they may lack. This can include procedures, equipment and methods to conduct research. Needed resources can also include access to bibliographic tools and materials that are simply unaffordable by institutions in the developing world.

Unfortunately, research indicates that forging collaborations between the developed and developing world can sometimes be difficult. For example, two-thirds of the 132 Filipinos who did graduate or post-doctoral work in Japan, Australia or the United States reported that they did not form international collaborations afterward (Ynalvez and Shrum 2009). Instead, the majority of these study-abroad graduate students collaborated with other Philippine scientists once they returned home (Ynalvez and Shrum 2009). In other words, their graduate study did not lead to collaborations with others outside their own country.

An alternative to formal research collaboration is participating in informal transnational networks. Through such networks, scholars gain "considerable material support… and a range of resources which are key to securing publication in Anglophone… contexts" (Lillis and Curry 2010, p. 85). Typically, informal networks are transient and reflect different academic-social status among the members: novice-to-expert or international-to-Anglophone. Informal relations are formed through conversations at conferences, funded invitations to Anglophone scholars to local academic events, and maintaining ties with fellow graduate students and doctoral advisors in developed countries. Informal networks provide support by alerting scholars outside the Anglophone countries of opportunities that they might otherwise be unaware of or unable to access alone (Box 12.1)

Box 12.1 Examples of Opportunities Made Available through Informal Networks

- An edited book seeks chapters and this information is often communicated simply by word of mouth.
- An introduction is made by a central network colleague to a book publisher or a journal editor with the consequence of improving the reception the Anglophone scholar receives and facilitating more favorable consideration than might otherwise occur.
- An informal read-through of a manuscript is performed before it is submitted for publication review in order to gain Anglophone comment on the text.
- An invitation to participate in international conference symposia involving several researchers on different projects is offered, raising the visibility of the non-Anglophone non-center researcher.

Different networks provide different resources, but crucially, networks "can provide information, opportunities, and support for English-medium text production, leading to publication (Lillis and Curry 2010, p. 86). Whether English-language publishing leads to participation in informal transnational networks, or vice versa, is not well documented. However, there is strong anecdotal evidence that participation in networks through conferences and so on, can facilitate English-language publishing.

The pressures and benefits to non-native-English speaking scholars to publish in English-language journals that carry impact factors have led to the creation of a service industry to support those scholars. In Europe, China and the United

States, profit-making companies have been established offering translation, editing, and revision services. Pre-packed webinars can be purchased on topics such as "Persuasive Writing for Scientists" that are marketed to scientists in the developing world. The non-profit mentoring organization AuthorAid attempts to match established and often retired scientists with international ones. As of July 2012, that organization reports more than 5000 scholars from 130 countries have registered to seek a mentor, and 700 scientists have offered to be mentors. At least one university in the United States, Texas A&M, offers a three-week intensive research writing course for international scholars that costs approximately $2,000 USD before travel or lodging expenses. In addition, there is a myriad of individuals who offer writing workshops, courses, and editing services to scientists who are hoping to improve their manuscripts to the satisfaction of journal editors of English-language journals.

In sum, non-native-English-speaking scientists face the additional burden of learning to write scientific papers in a language that is not their mother tongue. This burden is sometimes lessened to the degree that the scientist encounters no obstacle to their producing a publishable manuscript. In the vast majority of cases, scientists report that writing scientific manuscripts in a second language is more difficult than if they were to write in their mother tongue. There is evidence that, because English is the current lingua franca of science, there is a range of English that is accepted for publication purposes. However, producing linguistically adequate texts can be insufficient for obtaining publication in English-language journals, particularly if work is criticized for being too local or conversely, for over-generalizing. Collaborating in formal research arrangements or informal transnational networks can provide the access that international researchers need to obtain publication in the desired Anglophone, U.S./U.K journals.

References

Aydlini, E., & Mathews, J. (2000). Are the core and periphery irreconcilable? *International Studies Perspective, 1*(3), 289–303.

Ammon, U. (2001). Editor's preface. In U. Ammon (Ed.), *The dominance of English as a language of science* (pp. v-x). Berlin: Mouton de Gruyter.

Englander, K. (2009). Transformation of the identities of nonnative English speaking scientists. *Journal of Language, Identity and Education, 8*, 35–53.

Gosden, H. (1996). Verbal reports of Japanese novices' research writing practices in English. *Journal of Second Language Writing, 5*(2), 87–101.

Hanauer, D. I., & Englander, K. (2011). Quantifying the burden of writing research articles in a second language: Data from Mexican scientists. *Written Communication, 28*, 403–416.

Hanauer, D. I., & Englander, K. (2013). *Writing science in a second language*. West Lafayette, IN: Parlor Press.

Huang, J. C. (2010). Publishing and learning writing for publication in English: Perspectives of NNES PhD students in science. *English for Academic Purposes, 9*, 33–44.

Hwang, K. (2005). The inferior science and the dominant use of English in knowledge production: A case study of Korean science and technology. *Science Communication, 26*(4), 390–427.

Lillis, T., & Curry, M. J. (2010). *Academic writing in a global context: The politics and practices of publishing in English*. London: Routledge.
Matsumoto, K. (1995). Research paper writing strategies of professional Japanese EFL writers. *TESL Canada Journal, 13*(1), 17–27.
Moreno, A. I., Rey-Rocha, J., Burgess, S., López-Navarro, I., & Sachdev, I. (2012). Spanish researchers' perceived difficulty writing research articles for English-medium journals. *Ibérica, 24*, 157–184.
Pérez-Llantada, C. (2012). *Scientific discourse and the rhetoric of globalization*. London: Continuum.
Rozycki, W., & Johnson, N. H. (2013). Non-canonical grammar in Best Paper award winners in engineering. *English for Specific Purposes, 32*(3), 157–169.
Uzuner, S. (2008). Multilingual scholars' participation in core/global academic communities: A literature review. *English for Academic Purposes, 7*(4), 250–263.
Ynalvez, M. A., & Shrum, W. (2009). International graduate students training and scientific collaboration. *International Sociology, 24*, 870–901.

Chapter 13
Gatekeepers, Guardians and Allies

Abstract Journals editors, with the help of journal referees or reviewers, have a double role. They aim to weed out inadequate research and assure that new discoveries get disseminated to the community. Typical questions that reviewers consider are scientific integrity and the written presentation of the work. Reviewer comments can improve the final paper; however, not all reviewers are unbiased. Some reviewers respond to manuscripts that they deem to be insufficient with scorn as guardians of the discipline, and others take the role of ally with the author to bring the manuscript to publication.

Keywords Referees • Reviewers • Commentary • Rejection • Revise and resubmit • Editors • Journal • Bias • Acceptance

Journal editors call upon a network of fellow scientists and scholars to comment on a manuscript that is submitted before it is published. These members of the scientific community review the manuscript in order to determine its appropriateness for publication in that particular journal. The journal editor typically relies on those comments to determine whether the manuscript should be accepted 'as is' or it should be rejected; in between those two extremes are requests for revisions that may range from minor to major. Together, the reviewer's and editor's goal "is to ensure that inaccurate or sloppy research is weeded out" (Yaffe 2009, p. 1) but also to assure that "new discoveries … get disseminated to the scientific community as rapidly as possible" (Yaffe 2009, p. 1). Once a manuscript is accepted for publication, it is "up to the community at large to come to their own conclusions about the soundness of the research" (McGinty 1999, p. 138). Because journal editors and reviewers control access to publication, they are sometimes termed gatekeepers; that is, they prevent those who have not earned the right to enter from entering.

There are a number of typical questions that reviewers are asked to consider concerning a manuscript (Box 13.1). On the broadest level, the manuscript should be well presented to demonstrate that there is an important contribution based on appropriate methods that will be of interest to the journal readers.

K. Englander, *Writing and Publishing Science Research Papers in English*,
SpringerBriefs in Education, DOI: 10.1007/978-94-007-7714-9_13,

Box 13.1 Typical Questions Included on Reviewer Evaluation Forms

- Is the contribution new?
- Is the contribution significant?
- Is it suitable for the publication in the Journal?
- Is the organization acceptable?
- Do the methods and treatment of results conform to acceptable scientific standards?
- Are the conclusions firmly based in the data?
- Is the length of the paper satisfactory?
- Are all the illustrations required?
- Are all the figures and tables necessary?
- Are the figure legends and table captions adequate?
- Do the Title and Abstract clearly indicate the content of the paper?
- Are the references up to date?
- Is the paper excellent, good or poor?
  (Based on Cargill and O'Connor 2009, p. 16)

A fundamental tenet of the peer-review publishing process is that reviewers will offer a thoughtful and objective reading of the manuscript in order to bring sound science to the larger community. However, the neutrality of journal reviewers has been questioned in a number of studies. There is evidence that reviewers are influenced by the gender, Anglophone name and/or institutional affiliation of authors (Link 1998; Wenneras and Wold 1997; Kourilová 1996).

Reviewers tend to favor those whom they are most familiar with, and the overwhelming majority of reviewers of the sought-after English-medium journals are people in the U.S., U.K. and Canada. This preference for the known and familiar can put scientists who are affiliated with institutions outside of those countries at a disadvantage: their provenance can elicit a degree of skepticism from the reviewers that another author with a familiar provenance would not be subject to. One Korean physicist with more than 200 published papers noted that reviewers were more critical of his papers when he relocated from the U.S. to Korea (Cho 2009). He attributed the increased scrutiny to his physically being located outside an Anglophone, American institution.

One means of minimizing such possible bias is to recruit a broader group of reviewers from scientific associations and institutions around the world. Efforts are deliberately being made by journals such as *Neurology, Canadian Journal of Forestry* and the *International Journal of Cybercriminology* to recruit more international reviewers.

The comments that reviewers make about manuscripts have been studied (Belcher 2007; Gosden 2003; Fortanet 2008; Daniel 1993/2004). Generally, there are more negative or critical comments than positive comments in a reviewer's commentary. This may not be surprising as these reviews can be characterized as falling into one of three categories: "those that suggest improvements, those that find the paper to be… less than a publishable unit, and those that identify flaws beyond repair" (Ioannidis et al. 2010, p. 285). The rejection rate is high in many journals, often more than 70 %. When comments suggest ways to improve, it can be a favorable indication of the value of the paper.

Often scientists aim high; that is, they first submit to a high impact journal. If the paper is not accepted, they will submit to another journal of similarly high impact or of lesser impact. It seems that in some cases, an author will resubmit a previously rejected paper to the same journal, making a change to the title and other non-substantive revisions, but they do not address the initial reviewer comments (Ioannidis et al. 2010). This practice is not recommended and such duplicate submissions can earn one a blacklist within the publishing community.

Being rejected by a high impact journal does not mean that the paper will not be accepted in a different one, depending on the kinds of comments. Chemistry manuscripts that were rejected by a high impact journal due to negative comments from the reviewers about the "relevance of the contribution" and "design/conception" were *not* later published in another high impact journal (Bornmann et al. 2010). They were generally eventually published in a lower-impact journal. However, if the original high impact journal rejected the manuscript because of "reference to the literature and documentation", "method/statistics", "discussion of results" and "writing/presentation", the manuscript was statistically likely to be published later in a different high impact journal, or in a low impact one. In other words, these latter comments did not preclude the manuscripts being published in another high-impact journal. With this finding, readers of high impact journals can well assume that the papers published have acceptable design and implementation of the work, and its importance is well positioned.

Poor writing in itself is usually not a reason to reject a manuscript (Guardiano et al. 2007), although it can be. When manuscript acceptance rates are low, language can "be as good a reason as any" to reject a manuscript (Gosden 1992, p. 129). Intolerance for what is considered poor writing is made clear in an editorial in *The Scientist*, titled "No pardon for poor English in science" (Jaffe 2003). The former editor of the journal *Science* equates poor language skills with poor science, saying, "If you see people making multiple mistakes in spelling, syntax and semantics, you have to wonder whether when they did their science they were not also making similar errors of inattention" (cited in Gibbs 1995, p. 96). The prestigious journal *Applied Physics Letters* explicitly indicates that manuscripts can be rejected for language problems, as reported by Huang (2010).

The focus on language problems is certainly legitimate because a scientific paper needs to effectively communicate what was done, what was found and why it is important. Language problems are evident in manuscripts written by local Anglophone scientists as well as those written by international scientists

(Swales 2004). When the language problems are in a manuscript from a non-native-English speaker, some journal editors report that they provide extra attention and go "out of their way" to provide help (Flowerdew 2001, p. 129).

In three reported studies, about half to three-quarters of the international authors received a comment about insufficient English in their papers (Gosden 2003; Mungra and Webber 2010; Hanauer and Englander 2013). Even native-English-speaking authors who submit a paper from outside an Anglophone country have received reviewer comments telling them to have their paper "reviewed by a native speaker" (Belcher 2007; David Hanauer, personal communication, May 18, 2008). This suggests a predisposition on the part of the reviewer that an author who resides outside the Anglophone world could not submit a paper that is fluently written.

As comments on the non-native langauge elements is commonly noted by reviewers, an analysis was conducted of several sets of reviewer reports received by two Mexican scientists (Englander and López-Bonilla 2011). The reviewers' comments revealed what they considered to be "good science" and a "good paper". When the reviewer believed that the Mexican scientist shared those same criteria, the reviewer's attitude to the language difficulties in the manuscript was helpful. He played the role of an "ally" in assisting the author to improve the manuscript or a "ringmaster" in helping the author to acquire the necessary resources to improve the manuscript. However, when the reviewer felt that the author did not share the same beliefs about good science or a good paper, the reviewer played the role of "guardian" of the discipline, and the language difficulties were emphasized. "Although the reviewers commented on the *same manuscript, and all noted nonnative-English*, some severely criticized the language, and others offered linguistic assistance" (Hanauer and Englander 2013, p. 44, emphasis added). The reviewer's perception of whether or not the author shared the same attitude about science seemed to determine the decision to support or deny access to publication.

All in all, there is evidence that the peer review process does not guarantee that good work will get published and neither does it guarantee that bad work won't get published. Nonetheless, many scientists remark that peer review does make their papers better. Novice scientists learn from the comments of the reviewers about writing better manuscripts. Established scientists gain the opportunity to defend or limit their work before it goes to the wider scientific community. Peer review of manuscripts continues to be a collegial process that is considered central for scientific knowledge to progress.

References

Belcher, D. (2007). Seeking acceptance in an English-only research world. *Journal of Second Language Writing, 16*, 1–22.

Bornmann, L., Weymuth, C., & Daniel, H. D. (2010). A content analysis of referees' comments: How do comments on manuscripts rejected by a high-impact journal and later published in either a low- or high-impact journal differ? *Scientometrics, 83*, 493–506.

Cargill, M., & O'Connor, P. (2009). *Writing scientific research articles*. West Sussex: Wiley-Blackwell.

Cho, D. W. (2009). Science journal paper writing in an EFL context. *English for Specific Purposes, 28*, 230–239.

Daniel, H. D. (1993/2004). *Guardians of science: Fairness and reliability of peer review.* Weinheim: Wiley. Published online 2004 doi:10.1002/3527602208.

Englander, K., & López-Bonilla, G. (2011). Acknowledging or denying membership: Reviewers' responses to non-anglophone scientists' manuscripts. *Discourse Studies, 13*(5), 1–23.

Flowerdew, J. (2001). Attitudes of journal editors to nonnative speaker contributions. *TESOL Quarterly, 35*, 121–150.

Fortanet, I. (2008). Evaluative language in peer review referee reports. *English for Academic Purposes, 7*(1), 27–37.

Gibbs, W. (1995). Trends in scientific communication: Lost science in the third world. *Scientific American, 273*(2), 92-99.

Gosden, H. (1992). Research writing and NNSs: From the editors. *Journal of Second Language Writing, 1*(2), 123–139.

Gosden, H. (2003). 'Why not give us the full story?' Functions of referees' comments in peer review of scientific research papers. *English for Academic Purposes, 2*, 109–128.

Guardiano, C., Favilla, M., & Calaresu, E. (2007). Stereotypes about English as the language of science. *AILA Review, 20*, 28–52.

Hanauer, D. I., & Englander, K. (2013). *Writing science in a second language*. West Lafayette: Parlor Press.

Huang, J. C. (2010). Publishing and learning writing for publication in English: Perspectives of NNES PhD students in science. *English for Academic Purposes, 9*, 33–44.

Ioannidis, J. P., Tatsioni, A., & Karassa, F. (2010). Who is afraid of reviewer' comments? Or, why anything can be published and anything can be cited. *European Journal of Clinical Investigation, 40*, 285–287.

Jaffe, S. (2003). No pardon for poor English in science. *The Scientist, 17*(5), 44–45.

Kourilová, M. (1996). Interactive functions of language in peer reviews of medical papers written by non-native users of English. *UNESCO ALSED-LSP Newsletter, 19*(1), 4–21.

Link, A. M. (1998). U.S. and non-U.S. submissions: An analysis of reviewer bias. *Journal of the American Medical Association (JAMA), 280*(3), 246–247.

McGinty, S. (1999). *Gatekeepers of knowledge: Journal editors in the sciences and social sciences*. Westport: Bergin and Garvey.

Mungra, P., & Webber, P. (2010). Peer review process in medical research publications: Langauge and content comments. *English for Specific Purposes, 29*, 43–53.

Swales, J. M. (2004). *Research genres: Explorations and applications.* Cambridge: Cambridge University Press.

Wenneras, C., Wold, A. (1997). Nepotism and sexism in peer-review. *Nature, 387*, 341–343.

Yaffe, M. B. (2009). Re-reviewing peer review. *Science Signaling, 2*(85), 1–2.

Chapter 14
Afterword: Negotiating Research Article Writing and Publication

Abstract Never before have so many scientists around the world been so pressured to publish their work in English. At the same time, never before has there been such great opportunity for scientists around the world to communicate quickly and widely with colleagues who share an interest in scientific endeavors. English as a lingua franca in science opens up global conversations to native and non-native speakers alike. By publishing scientific work, knowledge capital and social capital are created. In contrast, when scientific investigation is not reported in a scientific paper, the work is lost. Successful publication sets a new foundation upon which further scientific discovery is possible.

Keywords Lingua franca • Social capital • Social activity • Knowledge capital • Relationships • Publishing • Role of science

This short volume has presented the research conducted in the last few decades about writing and publishing scientific articles in today's global world. The research comes from the fields of applied linguistics, rhetoric, sociology of science, history of science, and bibliometrics where the researchers use quantitative and qualitative approaches to understand the phenomenon.

The research makes clear that never before have so many scientists around the world been so pressured to publish their work in English. At the same time, never before has there been such great opportunity for scientists around the world to communicate quickly and widely with colleagues who share an interest in scientific endeavors. The use of English as a lingua franca in science opens up global scientific conversations to native and non-native speakers alike. This situation makes for a time of great excitement in science as formerly muted voices gain access to international conversations.

The scientific research paper is the most established medium for capturing and creating new knowledge. This is a great value in itself. The knowledge communicated

K. Englander, *Writing and Publishing Science Research Papers in English*,
SpringerBriefs in Education, DOI: 10.1007/978-94-007-7714-9_14,

in scientific papers helps explain the universe we live in and often helps us to make the world better. It is a kind of knowledge capital.

In addition to knowledge capital, scientific publishing also creates social capital. The individual scientist has an increase in prestige that can turn into funding opportunities, conference plenaries, and university promotions. As the prominence of the individual scientist rises, so does the prestige of the research team, the institution, and the nation. The pervasive system of metrics and rankings are a tangible form of reporting this social capital. As described in Part I of this book, this is the context in which science today takes place.

Scientific investigation which is not reported in a scientific paper is lost to the world. To prevent such loss, science education should inculcate a writing culture for university students and allow students to see for themselves the importance of creating appropriate scientific texts. Learning to write, think and work in the ways that are accepted in a discipline takes years, and so it must begin early. Students who do not speak English as their first language should have opportunities to gain scientific English skills. They should be permitted to ask the questions, develop the approaches and use the international sources that can broaden science for us all. Scientists everywhere need to be able to fulfill the stylistic and rhetorical demands of scientific writing. As shown in Part II of this book, those demands are high.

The scientific world is a social world. As discussed in Part III, relationships are crucial in science. Novices need to be apprenticed into not only the writing skills, but the social dimension of the scientific world: collaborations, conferences, networks and international alliances. Scientists who are located outside the predominantly English-speaking countries benefit greatly by informal networks and formal collaborations if they seek to publish in the English-language journals that are indexed in the major databases. They also may make decisions to form local networks to publish in their own language or in regional journals. In this way they may talk to a different audience than if they chose an English-language journal in the United States. Choosing the audience who is to be the scientist's readers is one of the many social elements of writing and publishing.

Bringing a paper into publication is a social activity. Editors and journal reviewers determine what they believe to be appropriate for their journal. If a paper is not outright rejected, a negotiation of improvements and changes occurs between the author and the editor. The scientist must consider what is being asked of him or her and justify the decisions made. Bringing the work to the larger scientific community is the immediate goal. But that goal is part of a much larger social project. By publishing their work, scientists set a new foundation upon which further discovery is made possible. This is a process by which we can truly, make the world a better place.

Kolibri 360 GmbH
Ferdinand-Jühlke-Straße 7
99095 Erfurt, Deutschland
produktsicherheit@kolibri360.de